D0856920

POCKET WORLD IN FIGURES
2009 EDITION

The
Economist

Pocket
World in
Figures
2009 Edition

THE ECONOMIST IN ASSOCIATION WITH
PROFILE BOOKS LTD

Published by Profile Books Ltd,
3A Exmouth House, Pine Street, London EC1R 0JH

This edition published by Profile Books in association with
The Economist, 2008

Copyright © The Economist Newspaper Ltd, 1991, 1992,
1993, 1994, 1995, 1996, 1997, 1998, 1999, 2000, 2001, 2002, 2003,
2004, 2005, 2006, 2007, 2008

Material researched and compiled by
Sophie Brown, Andrea Burgess, Ulrika Davies, Mark Doyle,
Ian Emery, Andrew Gilbert, Conrad Heine, Carol Howard,
Stella Jones, David McKelvey, Roxana Willis, Simon Wright

Typeset in Officina by MacGuru Ltd
info@macguru.org.uk

Printed in Italy by
Graphicom

A CIP catalogue record for this book is available
from the British Library

ISBN 978 1 84668 123 3

Contents

109 Part II   Country Profiles

Notes

This 2009 edition of *The Economist Pocket World in Figures*
includes new rankings on such things as privacy, gold
reserves, student performance, teenagers' behaviour, obesity,
mortgage debt, Oscars, press freedom, death row, slums and
various environmental measures. The world rankings consider
186 countries; all those with a population of at least 1m or a
GDP of at least $1bn; they are listed on pages 250–54. The
country profiles cover 67 major countries. Also included are
profiles of the euro area and the world. The extent and quality
of the statistics available varies from country to country. Every
care has been taken to specify the broad definitions on which
the data are based and to indicate cases where data quality or
technical difficulties are such that interpretation of the
figures is likely to be seriously affected. Nevertheless, figures
from individual countries may differ from standard
international statistical definitions. The term "country" can
also refer to territories or economic entities.

Some country definitions
Macedonia is officially known as the Former Yugoslav Republic
of Macedonia. Data for Cyprus normally refer to Greek Cyprus
only. Data for China do not include Hong Kong or Macau. For
countries such as Morocco they exclude disputed areas.
Congo-Kinshasa refers to the Democratic Republic of Congo,
formerly known as Zaire. Congo-Brazzaville refers to the other
Congo. Data for the EU refer to the 25 members prior to the
addition of Bulgaria and Romania on January 1 2007, unless
otherwise noted. Euro area data normally refer to the 12
members that had adopted the euro as at December 31 2006:
Austria, Belgium, France, Finland, Germany, Greece, Ireland,
Italy, Luxembourg, Netherlands, Portugal and Spain. For more
information about the euro area see page 248.

Statistical basis
The all-important factor in a book of this kind is to be able to
make reliable comparisons between countries. Although this
is never quite possible for the reasons stated above, the best
route, which this book takes, is to compare data for the same
year or period and to use actual, not estimated, figures
wherever possible. Where a country's data is excessively out of
date, it is excluded. The research for this edition of *The
Economist Pocket World in Figures* was carried out in 2008
using the latest available sources that present data on an
internationally comparable basis.

Data in the country profiles, unless otherwise indicated, refer to the year ending December 31 2006. Life expectancy , crude birth, death and fertility rates are based on 2005–10 averages; human development indices and energy data are for 2005; marriage and divorce data refer to the latest year for which figures are available. Employment, health and education data are for the latest year between 2000 and 2006.

Other definitions

Data shown in country profiles may not always be consistent with those shown in the world rankings because the definitions or years covered can differ.

Statistics for principal exports and principal imports are normally based on customs statistics. These are generally compiled on different definitions to the visible exports and imports figures shown in the balance of payments section.

Definitions of the statistics shown are given on the relevant page or in the glossary on page 248. Figures may not add exactly to totals, or percentages to 100, because of rounding or, in the case of GDP, statistical adjustment. Sums of money have generally been converted to US dollars at the official exchange rate ruling at the time to which the figures refer.

Energy consumption data are not always reliable, particularly for the major oil producing countries; consumption per head data may therefore be higher than in reality. Energy exports can exceed production and imports can exceed consumption if transit operations distort trade data or oil is imported for refining and re-exported.

Abbreviations

bn	billion (one thousand million)	m	million
EU	European Union	PPP	Purchasing power parity
kg	kilogram	TOE	tonnes of oil equivalent
km	kilometre	trn	trillion (one thousand billion)
GDP	Gross domestic product	...	not available
ha	hectare		

World rankings

Countries: natural facts

Countries: *the largest*[a]
'000 sq km

1	Russia	17,075		31	Tanzania	945
2	Canada	9,971		32	Nigeria	924
3	China	9,561		33	Venezuela	912
4	United States	9,373		34	Namibia	824
5	Brazil	8,512		35	Pakistan	804
6	Australia	7,682		36	Mozambique	799
7	India	3,287		37	Turkey	779
8	Argentina	2,767		38	Chile	757
9	Kazakhstan	2,717		39	Zambia	753
10	Sudan	2,506		40	Myanmar	677
11	Algeria	2,382		41	Afghanistan	652
12	Congo	2,345		42	Somalia	638
13	Saudi Arabia	2,200		43	Central African Rep	622
14	Greenland	2,176		44	Ukraine	604
15	Mexico	1,973		45	Madagascar	587
16	Indonesia	1,904		46	Kenya	583
17	Libya	1,760		47	Botswana	581
18	Iran	1,648		48	France	544
19	Mongolia	1,565		49	Yemen	528
20	Peru	1,285		50	Thailand	513
21	Chad	1,284		51	Spain	505
22	Niger	1,267		52	Turkmenistan	488
23	Angola	1,247		53	Cameroon	475
24	Mali	1,240		54	Papua New Guinea	463
25	South Africa	1,226		55	Sweden	450
26	Colombia	1,142		56	Morocco	447
27	Ethiopia	1,134			Uzbekistan	447
28	Bolivia	1,099		58	Iraq	438
29	Mauritania	1,031		59	Paraguay	407
30	Egypt	1,000		60	Zimbabwe	391

Mountains: *the highest*[b]

	Name	Location	Height (m)
1	Everest	Nepal-China	8,848
2	K2 (Godwin Austen)	Pakistan	8,611
3	Kangchenjunga	Nepal-Sikkim	8,586
4	Lhotse	Nepal-China	8,516
5	Makalu	Nepal-China	8,463
6	Cho Oyu	Nepal-China	8,201
7	Dhaulagiri	Nepal	8,167
8	Manaslu	Nepal	8,163
9	Nanga Parbat	Pakistan	8,125
10	Annapurna I	Nepal	8,091
11	Gasherbrum I	Pakistan-China	8,068
12	Broad Peak	Pakistan-China	8,047
13	Xixabangma (Gosainthan)	China	8,046
14	Gasherbrum II	Pakistan-China	8,035

a Includes freshwater.
b Includes separate peaks which are part of the same massif.

Rivers: *the longest*

Name	Location	Length (km)
1 Nile	Africa	6,695
2 Amazon	South America	6,516
3 Yangtze	Asia	6,380
4 Mississippi-Missouri system	North America	6,019
5 Ob'-Irtysh	Asia	5,570
6 Yenisey-Angara	Asia	5,550
7 Hwang He (Yellow)	Asia	5,464
8 Congo	Africa	4,667
9 Parana	South America	4,500
10 Mekong	Asia	4,425

Deserts: *the largest*

Name	Location	Area ('000 sq km)
1 Sahara	Northern Africa	8,600
2 Arabia	SW Asia	2,300
3 Gobi	Mongolia/China	1,166
4 Patagonian	Argentina	673
5 Great Victoria	W and S Australia	647
6 Great Basin	SW United States	492
7 Chihuahuan	N Mexico	450
8 Great Sandy	W Australia	400

Lakes: *the largest*

Name	Location	Area ('000 sq km)
1 Caspian Sea	Central Asia	371
2 Superior	Canada/US	82
3 Victoria	E Africa	69
4 Huron	Canada/US	60
5 Michigan	US	58
6 Tanganyika	E Africa	33
7 Baikal	Russia	32
8 Great Bear	Canada	31

Islands: *the largest*

Name	Location	Area ('000 sq km)
1 Greenland	North Atlantic Ocean	2,176
2 New Guinea	South-west Pacific Ocean	809
3 Borneo	Western Pacific Ocean	746
4 Madagascar	Indian Ocean	587
5 Baffin	North Atlantic Ocean	507
6 Sumatra	North-east Indian Ocean	474
7 Honshu	Sea of Japan-Pacific Ocean	227
8 Great Britain	Off coast of north-west Europe	218

Notes: Estimates of the lengths of rivers vary widely depending on eg, the path to take through a delta. The definition of a desert is normally a mean annual precipitation value equal to 250ml or less. Australia (7.69 sq km) is defined as a continent rather than an island.

Population: size and growth

Largest populations
Millions, 2006

1	China	1,323.6	34	Kenya	35.1
2	India	1,119.5	35	Algeria	33.4
3	United States	301.0	36	Canada	32.6
4	Indonesia	225.5	37	Morocco	31.9
5	Brazil	188.9	38	Afghanistan	31.1
6	Pakistan	161.2	39	Uganda	29.9
7	Bangladesh	144.4	40	Iraq	29.6
8	Russia	142.5	41	Peru	28.4
9	Nigeria	134.4	42	Nepal	27.7
10	Japan	128.2	43	Venezuela	27.2
11	Mexico	108.3	44	Uzbekistan	27.0
12	Vietnam	85.3	45	Malaysia	25.8
13	Philippines	84.5	46	Saudi Arabia	25.2
14	Germany	82.7	47	Taiwan	22.8
15	Ethiopia	79.3	48	Ghana	22.6
16	Egypt	75.4		North Korea	22.6
17	Turkey	74.2	50	Romania	21.6
18	Iran	70.3		Yemen	21.6
19	Thailand	64.8	52	Sri Lanka	20.9
20	France	60.7	53	Australia	20.4
21	United Kingdom	59.8	54	Mozambique	20.2
22	Congo-Kinshasa	59.3	55	Syria	19.5
23	Italy	58.1	56	Madagascar	19.1
24	Myanmar	51.0	57	Côte d'Ivoire	18.5
25	South Korea	48.0	58	Cameroon	16.6
26	South Africa	47.6	59	Chile	16.5
27	Colombia	46.3	60	Angola	16.4
28	Ukraine	46.0		Netherlands	16.4
29	Spain	43.4	62	Kazakhstan	14.8
30	Argentina	39.1	63	Cambodia	14.4
31	Tanzania	39.0		Niger	14.4
32	Poland	38.5	65	Mali	13.9
33	Sudan	37.0	66	Burkina Faso	13.6

Largest populations
Millions, 2050

1	India	1,592.7	15	Vietnam	116.7
2	China	1,392.3	16	Japan	112.2
3	United States	395.0	17	Russia	111.8
4	Pakistan	304.7	18	Iran	101.9
5	Indonesia	284.6	19	Turkey	101.2
6	Nigeria	258.1	20	Afghanistan	97.3
7	Brazil	253.1	21	Kenya	83.1
8	Bangladesh	242.9	22	Germany	78.8
9	Congo-Kinshasa	177.3	23	Thailand	74.6
10	Ethiopia	170.2	24	United Kingdom	67.1
11	Mexico	139.0	25	Tanzania	66.8
12	Philippines	127.1	26	Sudan	66.7
13	Uganda	126.9	27	Colombia	65.7
14	Egypt	125.9			

Fastest growing populations
Average annual % change, 2010–15

1	Niger	3.44	25	Equatorial Guinea	2.43
2	Timor-Leste	3.36	26	Guatemala	2.42
3	Burundi	3.22	27	Ethiopia	2.40
4	Uganda	3.21	28	Tanzania	2.36
5	Afghanistan	3.18	29	Gambia	2.32
6	Congo-Kinshasa	3.10	30	Sierra Leone	2.27
7	Liberia	3.09	31	Senegal	2.23
8	Guinea-Bissau	3.06	32	Mauritania	2.22
9	Mali	2.95	33	United Arab Emirates	2.13
10	Eritrea	2.94	34	Congo-Brazzaville	2.12
11	Yemen	2.90	35	Nigeria	2.09
12	West Bank and Gaza	2.87	36	Cape Verde	2.06
13	Angola	2.76	37	Saudi Arabia	2.05
	Benin	2.76	38	Kuwait	2.04
	Burkina Faso	2.76	39	Sudan	2.02
16	Somalia	2.74	40	Oman	1.95
17	Chad	2.73	41	Pakistan	1.90
18	Rwanda	2.72	42	Honduras	1.89
19	Guinea	2.65	43	Nepal	1.88
20	Iraq	2.60	44	Syria	1.85
21	Kenya	2.55	45	Côte d'Ivoire	1.84
22	Madagascar	2.48		Ghana	1.84
	Malawi	2.48		Zambia	1.84
24	Togo	2.44	48	Belize	1.83

Slowest growing populations
Average annual % change, 2010–15

1	Bulgaria	-0.80	23	Cuba	-0.01
2	Ukraine	-0.79		Italy	-0.01
3	Belarus	-0.57	25	Serbia	0.09
4	Russia	-0.56	26	Greece	0.10
5	Georgia	-0.53	27	Channel Islands	0.11
6	Romania	-0.53	28	Andorra	0.13
7	Latvia	-0.49	29	Denmark	0.14
8	Lithuania	-0.44		Martinique	0.14
9	Croatia	-0.34	31	Bermuda	0.15
	Moldova	-0.34		Montenegro	0.15
11	Estonia	-0.33		Netherlands	0.15
12	Hungary	-0.32		Portugal	0.15
13	Bosnia	-0.22	35	Taiwan	0.16
14	Japan	-0.18	36	Austria	0.17
15	Poland	-0.17		Belgium	0.17
16	Virgin Islands (US)	-0.15	38	South Korea	0.18
17	Germany	-0.13	39	Finland	0.23
18	Czech Republic	-0.09	40	Barbados	0.25
19	Slovenia	-0.08	41	North Korea	0.33
20	Armenia	-0.07	42	Uruguay	0.34
21	Macedonia	-0.04	43	Switzerland	0.35
22	Slovakia	-0.02	44	Trinidad & Tobago	0.37

Population: matters of breeding

Fertility rates, 2010–15

Highest av. no. of children per woman		Lowest av. no. of children per woman	
1	Niger	6.88	1	Macau	0.96
2	Guinea-Bissau	6.75	2	Hong Kong	0.99
3	Afghanistan	6.67	3	South Korea	1.21
4	Burundi	6.63	4	Belarus	1.23
5	Congo-Kinshasa	6.49	5	Ukraine	1.24
	Liberia	6.49	6	Poland	1.25
7	Sierra Leone	6.16	7	Bosnia	1.26
8	Mali	6.06	8	Japan	1.27
9	Angola	6.04	9	Slovakia	1.28
10	Timor-Leste	6.00	10	Singapore	1.29
	Uganda	6.00	11	Czech Republic	1.30
12	Chad	5.78	12	Lithuania	1.31
13	Somalia	5.61	13	Hungary	1.32
14	Burkina Faso	5.57		Romania	1.32
15	Rwanda	5.39	15	Latvia	1.33
16	Malawi	5.12		Slovenia	1.33
17	Equatorial Guinea	5.08	17	Bulgaria	1.34
18	Guinea	4.95	18	Russia	1.36
19	Yemen	4.93	19	Greece	1.38
20	Benin	4.92		Macedonia	1.38

Women[a] who use modern methods of contraception[b]

Highest, %		Lowest, %	
1	China	86	1	Somalia	1
2	Hong Kong	80	2	Chad	2
3	United Kingdom	79	3	Congo-Kinshasa	4
4	Finland	78		Guinea-Bissau	4
5	Netherlands	76		Sierra Leone	4
6	Australia	75	6	Angola	5
	Belgium	75		Eritrea	5
8	Canada	73		Mauritania	5
	Singapore	73		Niger	5
10	Costa Rica	72	10	Benin	6
	Cuba	72		Guinea	6
	Germany	72		Mali	6
	New Zealand	72	13	Central African Rep	7
14	Brazil	70		Côte d'Ivoire	7
	Thailand	70		Sudan	7
16	France	69	16	Albania	8
17	Colombia	68		Nigeria	8
	Hungary	68	18	Afghanistan	9
	Puerto Rico	68		Burkina Faso	9
	United States	68		Gambia, The	9
21	South Korea	67		Papua New Guinea	9
22	Nicaragua	66		Timor-Leste	9
	Vietnam	66		Togo	9
24	Austria	65			

a Married women aged 15–49. 2005 or latest available.
b Excludes traditional methods of contraception, such as the rhythm method.

Crude birth rates
Births per 1,000 population, 2007

Highest			Lowest	
1	Congo-Kinshasa	50	1 Germany	8
	Guinea-Bissau	50	Macau	8
	Liberia	50	3 Austria	9
4	Angola	49	Belarus	9
5	Mali	48	Bosnia	9
	Niger	48	Japan	9
	Sierra Leone	48	Lithuania	9
	Uganda	48	Slovenia	9
9	Afghanistan	47	South Korea	9
	Chad	47	Taiwan	9

Births per 1,000 women aged 15–19, 2005

1	Niger	244	14	Angola	138
2	Congo-Kinshasa	222	15	Nigeria	126
3	Liberia	219	16	Zambia	122
4	Uganda	203	17	Benin	120
5	Chad	189	18	Central African Rep	115
	Mali	189		Madagascar	115
7	Guinea-Bissau	188	20	Afghanistan	113
8	Guinea	176		Nicaragua	113
9	Timor-Leste	168	22	Gambia, The	109
10	Sierra Leone	160	23	Bangladesh	108
11	Burkina Faso	151	24	Côte d'Ivoire	107
12	Malawi	150		Guatemala	107
13	Congo-Brazzaville	143			

Sex ratio, males per 100 females, 2007

Highest			Lowest	
1	United Arab Emirates	210	1 Estonia	85
2	Qatar	203	Latvia	85
3	Kuwait	150	3 Russia	86
4	Bahrain	134	Ukraine	86
5	Oman	126	5 Armenia	87
6	Saudi Arabia	122	Belarus	87
7	Greenland	113	Lithuania	87
8	Bhutan	112	Netherlands Antilles	87
9	Afghanistan	107	9 Georgia	89
	Andorra	107	Lesotho	89
	Brunei	107	11 Martinique	90
	China	107	Virgin Islands (US)	90
	India	107	13 Hungary	91
	Libya	107	14 Aruba	92
15	Faroe Islands	106	Guadeloupe	92
	Jordan	106	Hong Kong	92
	Pakistan	106	Kazakhstan	92
18	Bangladesh	105	Macau	92
	French Polynesia	105	Moldova	92
20	Guam	104	Puerto Rico	92
	West Bank and Gaza	104		

Population: age

Median age[a]

Highest, 2007			Lowest, 2007		
1	Japan	42.9	1	Uganda	14.8
2	Italy	42.3	2	Niger	15.5
3	Germany	42.1	3	Mali	15.8
4	Andorra	42.0	4	Burkina Faso	16.2
5	Finland	40.9		Guinea-Bissau	16.2
6	Switzerland	40.8	6	Chad	16.3
7	Austria	40.6		Congo-Brazzaville	16.3
	Belgium	40.6		Congo-Kinshasa	16.3
	Bulgaria	40.6		Liberia	16.3
	Croatia	40.6		Malawi	16.3
11	Slovenia	40.2	11	Yemen	16.5
12	Sweden	40.1	12	Afghanistan	16.7
13	Channel Islands	39.7		Angola	16.7
	Greece	39.7		Zambia	16.7
15	Denmark	39.5	15	Burundi	17.0
	Latvia	39.5	16	West Bank and Gaza	17.1
	Portugal	39.5	17	Eritrea	17.4
18	France	39.3	18	Ethiopia	17.5
	Netherlands	39.3		Nigeria	17.5
20	Bermuda	39.0		Rwanda	17.5
	Czech Republic	39.0	21	Benin	17.6
	Ukraine	39.0		Equatorial Guinea	17.6
	United Kingdom	39.0	23	Mozambique	17.7
24	Estonia	38.9	24	Madagascar	17.8
	Hong Kong	38.9	25	Kenya	17.9
26	Hungary	38.8		Somalia	17.9
27	Canada	38.6		Togo	17.9
	Spain	38.6	28	Guinea	18.0
29	Norway	38.2	29	Central African Rep	18.1
30	Luxembourg	38.1		Guatemala	18.1
	Malta	38.1		Swaziland	18.1
32	Bosnia	38.0	32	Senegal	18.2
33	Belarus	37.8		Tanzania	18.2
	Lithuania	37.8	34	Mauritania	18.4
35	Singapore	37.5		Sierra Leone	18.4
36	Russia	37.3		Timor-Leste	18.4
37	Aruba	37.0	37	Côte d'Ivoire	18.5
38	Romania	36.7	38	Namibia	18.6
39	Australia	36.6	39	Zimbabwe	18.7
	Macau	36.6	40	Cameroon	18.8
41	Poland	36.5	41	Iraq	19.1
	Serbia	36.5		Laos	19.1
43	Martinique	36.4	43	Lesotho	19.2
44	Netherlands Antilles	36.2	44	Cape Verde	19.3
45	United States	36.1		Tajikistan	19.3
46	New Zealand	35.8	46	Gabon	19.4
47	Cuba	35.6	47	Nicaragua	19.7
	Slovakia	35.6		Papua New Guinea	19.7

a Age at which there are an equal number of people above and below.

Ageing index
Number of people aged 60 or over for every 100 people under 15, 2007

	Highest			Lowest	
1	Japan	201.0	1	Niger	6.6
2	Italy	189.8	2	Uganda	7.4
3	Germany	182.3	3	Liberia	7.7
4	Bulgaria	172.5	4	United Arab Emirates	7.9
5	Greece	166.0	5	Yemen	8.0
6	Latvia	164.4	6	Angola	8.5
7	Austria	156.1	7	Mali	8.6
8	Slovenia	155.9	8	Burkina Faso	8.9
9	Czech Republic	150.7		Congo-Kinshasa	8.9
10	Croatia	150.0	10	Eritrea	9.0
11	Ukraine	149.5	11	Burundi	9.2
12	Spain	149.2	12	Rwanda	9.3
13	Estonia	148.3	13	Congo-Brazzaville	9.4
14	Portugal	144.3		Somalia	9.4
15	Switzerland	142.9	15	Afghanistan	9.5
16	Sweden	142.8	16	Kenya	9.6
17	Hungary	140.1	17	Chad	9.7
18	Belgium	139.2		Guinea-Bissau	9.7
19	Finland	134.3	19	West Bank and Gaza	9.8
	Lithuania	134.3	20	Malawi	9.9
21	Romania	130.3	21	Benin	10.0
22	Belarus	126.9	22	Papua New Guinea	10.2
23	Channel Islands	126.3	23	Zambia	10.3
24	Bosnia	125.5	24	Ethiopia	10.7
25	United Kingdom	124.7	25	Nigeria	11.0
26	France	121.1	26	Madagascar	11.1
27	Malta	119.9	27	Iraq	11.2
28	Denmark	117.9	28	Togo	11.4
29	Hong Kong	116.1	29	Senegal	11.8
30	Russia	114.0		Timor-Leste	11.8
31	Netherlands	112.6	31	Mozambique	12.0
32	Poland	112.3	32	Mauritania	12.3
33	Canada	110.3	33	Tanzania	12.4
34	Norway	108.0	34	Qatar	12.6
35	Slovakia	106.1	35	Sierra Leone	12.8
36	Serbia	105.4	36	Saudi Arabia	12.9
37	Georgia	101.2	37	Côte d'Ivoire	13.0
38	Luxembourg	98.6		Oman	13.0
39	Australia	95.0	39	Equatorial Guinea	13.2
40	Cyprus	91.8		Guinea	13.2
41	Cuba	87.4	41	Laos	13.3
42	Macedonia	85.5		Nicaragua	13.3
43	Martinique	84.3		Syria	13.3
44	New Zealand	84.0	44	Tajikistan	13.4
45	United States	83.9	45	Cape Verde	13.5
46	South Korea	83.4	46	Namibia	13.8
47	Moldova	81.6	47	Cameroon	14.0
48	Puerto Rico	81.1	48	Kuwait	14.1
49	Macau	79.7	49	Central African Republic	14.2
50	Virgin Islands (US)	79.5		Zimbabwe	14.2

City living

Urban population
Highest, %, 2005

1	Bermuda	100.0		15	Australia	92.7
	Cayman Islands	100.0		16	Luxembourg	92.4
	Hong Kong	100.0		17	Qatar	92.3
	Singapore	100.0			Réunion	92.3
5	Guadeloupe	99.8		19	Malta	92.1
6	Macau	98.9		20	Israel	91.7
7	Puerto Rico	97.5		21	Andorra	91.3
8	Belgium	97.3		22	Argentina	90.6
9	Kuwait	96.4		23	Bahrain	90.2
10	Martinique	96.2		24	Bahamas	90.0
11	Virgin Islands (US)	94.1		25	United Kingdom	89.2
12	Guam	94.0		26	Germany	88.5
13	Iceland	93.0			Saudi Arabia	88.5
	Uruguay	93.0				

Urban population
Lowest, %, 2005

1	Timor-Leste	7.8		14	Laos	21.6
2	Bhutan	9.1		15	Rwanda	21.8
3	Burundi	10.6		16	Niger	23.3
4	Uganda	12.4		17	Swaziland	23.9
5	Papua New Guinea	13.2		18	Tajikistan	24.2
6	Nepal	15.8		19	Afghanistan	24.3
7	Ethiopia	16.2		20	Bangladesh	25.0
8	Malawi	17.2		21	Chad	25.8
9	Lesotho	18.2		22	Gambia	26.1
10	Burkina Faso	18.6		23	Yemen	26.3
11	Cambodia	19.7		24	Vietnam	26.7
12	Eritrea	20.8		25	Madagascar	27.0
13	Sri Lanka	21.0		26	India	28.7

Urban population
Biggest increase, %, 2005

1	Rwanda	11.6		19	Timor-Leste	4.8
2	Burundi	6.5		20	Equatorial Guinea	4.7
3	Bhutan	6.3		21	Chad	4.6
4	Niger	6.1			Laos	4.6
5	Afghanistan	6.0			Malawi	4.6
6	Eritrea	5.8			Sudan	4.6
7	Sierra Leone	5.7		25	Benin	4.4
	Somalia	5.7			Congo-Kinshasa	4.4
9	Cambodia	5.5			Kenya	4.4
10	Angola	5.4			Nigeria	4.4
	Guinea-Bissau	5.4		29	Ethiopia	4.1
12	Liberia	5.3			West Bank and Gaza	4.1
13	Mali	5.2		31	Togo	4.0
	Nepal	5.2		32	Indonesia	3.9
15	Mauritania	5.1			Senegal	3.9
	Mozambique	5.1			Uganda	3.9
17	Burkina Faso	5.0		35	Guinea	3.8
18	Tanzania	4.9				

Biggest cities[a]
Population m, 2005

1	Tokyo, Japan	35.3		Jakarta, Indonesia	8.8
2	Mexico City, Mexico	18.7		Lagos, Nigeria	8.8
	New York, US	18.7	25	London, UK	8.5
4	São Paulo, Brazil	18.3	26	Guangzhou, China	8.4
5	Mumbai, India	18.2	27	Lima, Peru	7.7
6	Delhi, India	15.1		Tehran, Iran	7.7
7	Shanghai, China	14.5	29	Bogotá, Colombia	7.4
8	Kolkata, India	14.3	30	Shenzhen, China	7.2
9	Buenos Aires, Argentina	12.6	31	Hong Kong	7.1
	Dhaka, Bangladesh	12.6		Kinshasa, Congo	7.1
11	Los Angeles, US	12.3		Tianjin, China	7.1
12	Karachi, Pakistan	11.6		Wuhan, China	7.1
13	Cairo, Egypt	11.5	35	Chennai, India	6.9
	Rio de Janeiro, Brazil	11.5	36	Bangkok, Thailand	6.6
15	Osaka, Japan	11.3	37	Bangalore, India	6.5
16	Manila, Philippines	10.8	38	Chongqing, China	6.4
17	Beijing, China	10.7	39	Lahore, Pakistan	6.3
18	Moscow, Russia	10.4	40	Hyderabad, India	6.1
19	Paris, France	9.9	41	Santiago, Chile	5.6
20	Seoul, South Korea	9.8	42	Madrid, Spain	5.4
21	Istanbul, Turkey	9.7		Miami, US	5.4
22	Chicago, US	8.8		Philadelphia, US	5.4

Proportion of a country's population residing in a single city[b]
%, 2005

1	Hong Kong	100.0	20	Lima, Peru	28.4
2	Singapore	100.0	21	San José, Costa Rica	28.1
3	Kuwait City, Kuwait	69.9	22	Tokyo, Japan	27.6
4	San Juan, Puerto Rico	66.0	23	Vienna, Austria	27.3
5	Montevideo, Uruguay	45.9	24	Lisbon, Portugal	26.2
6	Tel Aviv, Israel	45.0	25	Dublin, Ireland	25.0
7	Beirut, Lebanon	44.3	26	Tbilisi, Georgia	24.4
8	Panama City, Panama	37.6	27	Baku, Azerbaijan	22.4
9	Yerevan, Armenia	36.5	28	Luanda, Angola	22.0
10	Tripoli, Libya	35.4	29	Santo Domingo, Dom. Rep.	21.8
11	Santiago, Chile	34.4	30	Lomé, Togo	21.1
12	Brazzaville, Congo-Brazz.	33.7	31	Sydney, Australia	21.0
13	Ulan Bator, Mongolia	33.2	32	Helsinki, Finland	20.9
14	Monrovia, Liberia	33.1	33	Dakar, Senegal	20.7
15	Buenos Aires, Argentina	32.4	34	San Salvador, El Salvador	20.6
16	Dubai, United Arab Em.	31.0	35	Seoul, South Korea	20.5
17	Asunción, Paraguay	29.8	36	Port-au-Prince, Haiti	20.3
18	Athens, Greece	29.1	37	Copenhagen, Denmark	20.0
19	Auckland, New Zealand	29.0	38	Havana, Cuba	19.4

a Urban agglomerations. Data may change from year-to-year based on reassessments of agglomeration boundaries.
b Urban agglomerations over 750,000.

Fastest growing cities[a]
Average annual growth, 2005–10, %

1	Naypyitaw, Myanmar	57.8	31	Karaj, Iran	3.7
2	Abuja, Nigeria	8.3		Kumasi, Ghana	3.7
3	Al-Hudaydah, Yemen	7.0		Lagos, Nigeria	3.7
4	Taizz, Yemen	6.3		Mombasa, Kenya	3.7
5	Luanda, Angola	6.0		Yaoundé, Cameroon	3.7
6	Huambo, Angola	5.8	36	Chittagong, Bangladesh	3.6
7	Klang, Malaysia	5.7	37	Addis Ababa, Ethiopia	3.5
8	Sanaa, Yemen	5.3		Douala, Cameroon	3.5
9	Kinshasa, Congo-Kins.	4.8		Dubai, United Arab Emir.	3.5
	Lomé, Togo	4.8	40	Ghaziabad, India	3.4
	Ouagadougou, Burk. Faso	4.8		Zamboanga, Philippines	3.4
12	Kathmandu, Nepal	4.7	42	Antananarivo, Madagas.	3.3
13	Kabul, Afghanistan	4.6		Brasilia, Brazil	3.3
	Kananga, Congo-Kins.	4.6		Dhaka, Bangladesh	3.3
	Mbuji-Mayi, Congo-Kins.	4.6		Nanchang, China	3.3
16	Johore Bharu, Malaysia	4.5	46	Abidjan, Côte d'Ivoire	3.2
	N'Djaména, Chad	4.5		Accra, Ghana	3.2
18	Bamako, Mali	4.4		Jinxi, China	3.2
19	Brazzaville, Congo-Braz.	4.3		Mosul, Iraq	3.2
	Dar es Salaam, Tanzania	4.3		Port-au-Prince, Haiti	3.2
	Lubumbashi, Congo-Kins.	4.3		Santa Cruz, Boliva	3.2
22	Florianópolis, Brazil	4.0		Shantou, China	3.2
	Kigali, Rwanda	4.0		Surat, India	3.2
	Kuwait City, Kuwait	4.0		Yuci, China	3.2
25	Maputo, Mozambique	3.9	55	Conarky, Guinea	3.1
	Niamey, Niger	3.9		Cotonou, Benin	3.1
	Pekan Baru, Indonesia	3.9		Faridabad, India	3.1
28	Kampala, Uganda	3.8		Xuzhou, China	3.1
	Nairobi, Kenya	3.8	59	Islamabad, Pakistan	3.0
	Phnom Penh, Cambodia	3.8			

Slowest growing cities[a]
Average annual growth, 2005–10, %

1	Lodz, Poland	-0.67		New Orleans, US	-0.27
	Ufa, Russia	-0.67	17	Nizhniy Nov., Russia	-0.26
3	Busan, South Korea	-0.64	18	Omsk, Russia	-0.20
4	Yueyang, China	-0.63		Dnipropetrovsk, Ukraine	-0.20
5	Odessa, Ukraine	-0.53	20	Belgrade, Serbia	-0.18
6	Saratov, Russia	-0.52	21	Rostov-on-Don, Russia	-0.17
7	Zaporizhzhya, Ukraine	-0.48	22	Pingdingshan, China	-0.16
8	Volvograd, Russia	-0.43		Turin, Italy	-0.16
9	Donetsk, Ukraine	-0.39		Voronezh, Russia	-0.16
10	Saint Petersburg, Russia	-0.36	25	Seoul, South Korea	-0.13
11	Budapest, Hungary	-0.35	26	Chelyabinsk, Russia	-0.11
	Novosibirsk, Russia	-0.35	27	Rome, Italy	-0.10
13	Samara, Russia	-0.34	28	Milan Italy	-0.09
14	Montevideo, Uruguay	-0.28		Kharkiv, Ukraine	-0.09
15	Havana, Cuba	-0.27	30	Daegu, South Korea	-0.08
			31	Prague, Czech Rep.	-0.07

a With populations over 750,000.

Refugees and asylum seekers[a]

Largest refugee nationalities
'000, 2006

1	Afghanistan	2,107.5	11	Myanmar	202.8
2	Iraq	1,450.9	12	Bosnia	200.0
3	Sudan	868.3	13	Eritrea	193.7
4	Somalia	464.0	14	Serbia	174.0
5	Congo-Kinshasa	401.9	15	Liberia	160.6
6	Burundi	396.5	16	Russia	159.4
7	Vietnam	374.3	17	China	140.6
8	West Bank and Gaza	334.1	18	Azerbaijan	126.1
9	Turkey	227.2	19	Sri Lanka	117.0
10	Angola	206.5	20	Bhutan	108.1

Countries with largest refugee populations
'000, 2006

1	Pakistan	1,044.5	11	Kenya	272.5
2	Iran	968.4	12	Uganda	272.0
3	United States	843.5	13	Saudi Arabia	240.8
4	Syria	702.2	14	Congo-Kinshasa	208.4
5	Germany	605.4	15	Sudan	196.2
6	Jordan	500.2	16	India	158.4
7	Tanzania	485.3	17	Canada	151.8
8	United Kingdom	301.6	18	France	145.0
9	China	301.0	19	Thailand	133.1
10	Chad	286.7	20	Nepal	128.2

Origin of asylum applications to indust. countries
'000, 2006

1	Iraq	22.2	11	Mexico	6.8
2	China	18.3	12	Bangladesh	6.4
3	Russia	15.7	13	Eritrea	6.3
	Serbia	15.7		Nigeria	6.3
5	Turkey	8.7	15	Colombia	6.1
6	Afghanistan	8.4	16	Sri Lanka	5.7
7	Iran	8.3	17	Congo-Kinshasa	5.5
8	Pakistan	7.6	18	India	5.4
9	Somalia	7.3	19	Armenia	4.2
10	Haiti	7.0	20	Syria	3.7

Asylum applications in industrialised countries
'000, 2006

1	United States	51.5	10	Belgium	11.6
2	France	30.7	11	Switzerland	10.6
3	United Kingdom	27.9	12	Italy	10.1
4	Sweden	24.3	13	Norway	5.3
5	Canada	22.9		Spain	5.3
6	Germany	21.0	15	Cyprus	4.6
7	Netherlands	14.5		Turkey	4.6
8	Austria	13.4	17	Ireland	4.3
9	Greece	12.3	18	Poland	4.2

a As reported by UNHCR.

The world economy

Biggest economies
GDP, $bn, 2006

1	United States	13,164		24	Poland	339
2	Japan	4,368		25	Norway	335
3	Germany	2,897		26	Austria	322
4	China	2,645		27	Greece	308
5	United Kingdom	2,377		28	Denmark	275
6	France[a]	2,248		29	South Africa	255
7	Italy	1,851		30	Ireland	220
8	Canada	1,272		31	Iran	218
9	Spain	1,225		32	Argentina	214
10	Brazil	1,067		33	Finland	211
11	Russia	987		34	Thailand	206
12	India	912		35	Portugal	195
13	South Korea	888		36	Hong Kong	190
14	Mexico	839		37	Venezuela	182
15	Australia	781		38	Colombia	153
16	Netherlands	662		39	Malaysia	151
17	Turkey	403		40	Chile	146
18	Belgium	394		41	Czech Republic	143
19	Sweden	384		42	Israel	140
20	Switzerland	380		43	Singapore	132
21	Indonesia	365		44	United Arab Emirates[b]	130
	Taiwan	365		45	Pakistan	127
23	Saudi Arabia	349		46	Romania	122

Biggest economies by purchasing power
GDP PPP, $bn, 2006

1	United States	13,164		23	Thailand	482
2	China	6,092		24	Argentina	469
3	Japan	4,081		25	South Africa	431
4	India	2,740		26	Pakistan	375
5	Germany	2,663		27	Egypt	367
6	United Kingdom	2,003		28	Belgium	354
7	France	1,960		29	Greece	350
8	Russia	1,869		30	Malaysia	327
9	Italy	1,710		31	Sweden	311
10	Brazil	1,694		32	Austria	299
11	Mexico	1,269			Venezuela	299
12	Spain	1,264		34	Colombia	291
13	Canada	1,199			Ukraine	291
14	South Korea	1,113		36	Switzerland	279
15	Indonesia	770		37	Philippines	272
16	Australia	736		38	Hong Kong	268
17	Taiwan	722		39	Nigeria	233
18	Iran	694			Norway	233
19	Turkey	614		41	Czech Republic	227
20	Netherlands	597		42	Romania	225
21	Poland	566		43	Portugal	220
22	Saudi Arabia	528		44	Chile	214

Note: For a list of 186 countries with their GDPs, see pages 250–254.
a Includes overseas departments. b 2005

Regional GDP

$bn, 2007		*% annual growth 2002–07*	
World	54,312	World	4.6
Advanced economies	39,131	Advanced economies	2.7
G7	30,419	G7	2.4
Euro area (15)	12,158	Euro area (15)	2.0
Asia[a]	5,724	Asia[a]	9.0
Latin America	3,450	Latin America	4.8
Eastern Europe[b]	3,527	Eastern Europe[b]	6.9
Middle East	1,387	Middle East	6.0
Africa	1,092	Africa	5.9

Regional purchasing power

GDP, % of total, 2007		*$ per head, 2007*	
World	100.0	World	9,730
Advanced economies	56.4	Advanced economies	35,780
G7	43.5	G7	37,380
Euro area (15)	16.1	Euro area (15)	32,940
Asia[a]	20.1	Asia[a]	3,840
Latin America	8.3	Latin America	9,760
Eastern Europe[b]	8.6	Eastern Europe[b]	11,700
Middle East	3.8	Middle East	10,350
Africa	3.0	Africa	2,420

Regional population

% of total (6.7bn), 2007		*No. of countries[c], 2007*	
Advanced economies	15.3	Advanced economies	31
G7	11.3	G7	7
Euro area (15)	4.7	Euro area (15)	15
Asia[a]	52.6	Asia[a]	23
Latin America	8.6	Latin America	32
Eastern Europe[b]	7.1	Eastern Europe[b]	26
Middle East	3.7	Middle East	13
Africa	12.8	Africa	47

Regional international trade

Exports of goods & services, % of tot., 2007		*Current account balances, $bn, 2007*	
Advanced economies	66.4	Advanced economies	-463
G7	38.4	G7	-544
Euro area (15)	29.5	Euro area (15)	-30
Asia[a]	13.2	Asia[a]	384
Latin America	5.1	Latin America	16
Eastern Europe[b]	8.0	Eastern Europe[b]	-45
Middle East	4.7	Middle East	275
Africa	2.5	Africa	2

a Excludes Hong Kong, Japan, Singapore, South Korea and Taiwan.
b Includes Russia, other CIS and Turkey.
c IMF definition.

Living standards

Highest GDP per head
$, 2006

1	Luxembourg	87,490	36	Guadeloupe[b]	23,880
2	Norway	72,810	37	Cyprus	23,430
3	Bermuda[ac]	68,180	38	Aruba[ab]	22,810
4	Qatar	59,570		Martinique[b]	22,810
5	Iceland	54,400	40	Bahrain[b]	22,620
6	Ireland	52,410	41	Israel	20,660
7	Switzerland	52,110	42	Bahamas[a]	20,130
8	Denmark	50,990	43	Puerto Rico[a]	18,720
9	United States	43,730	44	Slovenia	18,650
10	Channel Islands[ab]	49,680	45	Portugal	18,550
11	Cayman Islands[ac]	43,090	46	South Korea	18,500
12	Sweden	42,180	47	Réunion[b]	18,230
13	Netherlands	40,380	48	Taiwan	15,990
14	Finland	39,750	49	Malta	15,940
	United Kingdom	39,750	50	Equatorial Guinea	15,860
16	Austria	39,270	51	New Caledonia[ad]	15,070
17	Andorra[ab]	39,010	52	Guam[ab]	14,620
	Canada	39,010	53	Virgin Islands (US)[ac]	14,470
19	Australia	38,260	54	Czech Republic	14,020
20	Belgium	37,890	55	Trinidad & Tobago	13,950
21	France	37,040	56	Saudi Arabia	13,850
22	Kuwait	36,460	57	French Polynesia[ae]	13,620
23	Faroe Islands[ab]	36,170	58	Estonia	12,620
24	Germany	35,030	59	Netherlands Antilles[ac]	12,610
25	Japan	34,080	60	Barbados	12,340
26	Italy	31,860	61	Oman[b]	11,860
27	Brunei	31,420	62	Hungary	11,180
28	Macau	31,340	63	Slovakia	10,190
29	Greenland[ab]	30,360	64	Croatia	9,330
30	Singapore	30,040	65	Chile	8,840
31	Spain	28,220	66	Poland	8,800
32	Greece	27,790	67	Latvia	8,750
33	United Arab Emirates[b]	27,600		Lithuania	8,750
34	Hong Kong	26,730	69	Libya	8,390
35	New Zealand	25,490	70	Mexico	7,750

Lowest GDP per head
$, 2006

1	Burundi	120	11	Zimbabwe[b]	260
2	Congo-Kinshasa	140	12	Afghanistan	270
3	Ethiopia	170		Rwanda	270
4	Guinea-Bissau	190	14	Madagascar	290
	Liberia	190		Somalia[a]	290
6	Myanmar	230	16	Gambia, The	320
7	Eritrea	240		Nepal	320
	Malawi	240		Uganda	320
9	Niger	250	19	Tanzania	330
	Sierra Leone	250	20	Mozambique	340

a Estimate. b 2005 c 2004 d 2003 e 2002

Highest purchasing power
GDP per head in PPP (USA = 100), 2006

1	Luxembourg	172.0	36	Spain	65.2
2	Qatar[b]	160.8	37	Equatorial Guinea	61.8
3	Bermuda[ac]	159.0	38	Cyprus	58.9
4	Channel Islands[b]	117.9	39	New Zealand	58.0
5	Norway	113.9	40	Slovenia	55.4
6	Brunei	113.5	41	Israel	54.8
7	Singapore	101.7	42	Bahamas[b]	52.5
8	Macau	100.0	43	South Korea	52.3
	United States	100.0	44	Saudi Arabia	50.7
10	Cayman Islands[ac]	99.6	45	Czech Republic	50.3
11	Kuwait[b]	99.1	46	Aruba[ac]	49.6
12	Ireland	91.6	47	Malta	49.4
13	Hong Kong	88.8	48	Martinique[b]	47.9
14	Andorra[ab]	88.2	49	Portugal	47.3
15	Switzerland	84.6	50	Oman[b]	46.3
16	Iceland	84.0	51	Greenland[ae]	45.5
17	Canada	83.5	52	Guadeloupe[b]	44.7
18	Netherlands	83.2	53	Puerto Rico[a]	43.4
19	Austria	82.0	54	Estonia	43.1
20	Denmark	81.2	55	Hungary	41.6
21	Australia	80.8	56	Slovak Republic	40.3
22	Isle of Man[ab]	79.6		Trinidad & Tobago	40.3
23	Sweden	77.8	58	French Polynesia[ad]	39.8
24	Belgium	76.3	59	Réunion[b]	39.0
25	United Arab Emirates[b]	76.2	60	Netherlands Antilles[ac]	36.4
26	Bahrain[b]	76.1	61	Barbados[b]	36.2
27	United Kingdom	75.3	62	Lithuania	35.8
28	Finland	75.1	63	Latvia	34.9
29	Germany	73.5	64	Guam[ab]	34.1
30	France	72.8		New Caledonia[ad]	34.1
31	Japan	72.7	66	Poland	33.7
32	Taiwan[a]	72.3	67	Virgin Islands (US)[ac]	33.0
33	Greece	71.4	68	Croatia	32.5
34	Faroe Islands[ae]	70.5	69	Gabon	32.3
35	Italy	66.1	70	Russia	29.8

Lowest purchasing power
GDP per head in PPP (USA = 100), 2006

1	Congo-Kinshasa	0.6		Central African Rep	1.6
2	Burundi	0.8		Malawi	1.6
	Liberia	0.8	12	Rwanda	1.7
4	Guinea-Bissau	1.1		Mozambique	1.7
5	Somalia[a]	1.4	14	Togo	1.8
	Niger	1.4	15	Myanmar[b]	1.9
	Sierra Leone	1.4	16	Madagascar	2.0
	Ethiopia	1.4		Uganda	2.0
9	Eritrea	1.6			

Note: for definition of purchasing power parity see page 249.
a Estimate.   b 2005   c 2004   d 2003   e 2001

The quality of life

Human development index[a]

Highest, 2005

1	Iceland	96.8		31	Barbados	89.2
	Norway	96.8		32	Czech Republic	89.1
3	Australia	96.2			Kuwait	89.1
4	Canada	96.1		34	Malta	87.8
5	Ireland	95.9		35	Qatar	87.5
6	Sweden	95.6		36	Hungary	87.4
7	Switzerland	95.5		37	Poland	87.0
8	Japan	95.3		38	Argentina	86.9
	Netherlands	95.3		39	United Arab Emirates	86.8
10	Finland	95.2		40	Chile	86.7
	France	95.2		41	Bahrain	86.6
12	United States	95.1		42	Slovakia	86.3
13	Denmark	94.9		43	Lithuania	86.2
	Spain	94.9		44	Estonia	86.0
15	Austria	94.8		45	Latvia	85.5
16	Belgium	94.6		46	Uruguay	85.2
	United Kingdom	94.6		47	Croatia	85.0
18	Luxembourg	94.4		48	Costa Rica	84.6
19	New Zealand	94.3		49	Bahamas	84.5
20	Italy	94.1		50	Cuba	83.8
21	Hong Kong	93.7		51	Mexico	82.9
22	Germany	93.5		52	Bulgaria	82.4
23	Israel	93.2		53	Libya	81.8
24	Greece	92.6		54	Oman	81.4
25	Singapore	92.2			Trinidad & Tobago	81.4
26	South Korea	92.1		56	Romania	81.3
27	Slovenia	91.7		57	Panama	81.2
28	Cyprus	90.3			Saudi Arabia	81.2
29	Portugal	89.7		59	Malaysia	81.1
30	Brunei	89.4				

Human development index[a]

Lowest, 2005

1	Sierra Leone	33.6		10	Congo-Kinshasa	41.1
2	Burkina Faso	37.0		11	Burundi	41.3
3	Guinea-Bissau	37.4		12	Côte d'Ivoire	43.2
	Niger	37.4		13	Zambia	43.4
5	Mali	38.0		14	Benin	43.7
6	Central African Rep	38.4			Malawi	43.7
	Mozambique	38.4		16	Angola	44.6
8	Chad	38.8		17	Rwanda	45.2
9	Ethiopia	40.6		18	Guinea	45.6

a GDP or GDP per head is often taken as a measure of how developed a country is, but its usefulness is limited as it refers only to economic welfare. In 1990 the UN Development Programme published its first estimate of a Human Development Index, which combined statistics on two other indicators – adult literacy and life expectancy – with income levels to give a better, though still far from perfect, indicator of human development. In 1991 average years of schooling was combined with adult literacy to give a knowledge variable. The HDI is shown here scaled from 0 to 100; countries scoring over 80 are considered to have high human development, those scoring from 50 to 79 medium and those under 50 low.

Economic freedom index[a], 2008

1	Hong Kong	90.3	21	Barbados	71.3
2	Singapore	87.4		Cyprus	71.3
3	Ireland	82.4	23	Germany	71.2
4	Australia	82.0	24	Bahamas	71.1
5	United States	80.6	25	Taiwan	71.0
6	Canada	80.2	26	Lithuania	70.8
	New Zealand	80.2	27	Sweden	70.4
8	Chile	79.8	28	Armenia	70.3
9	Switzerland	79.7	29	Trinidad & Tobago	70.2
10	United Kingdom	79.5	30	Austria	70.0
11	Denmark	79.2	31	Spain	69.7
12	Estonia	77.8	32	El Salvador	69.2
13	Netherlands	76.8		Georgia	69.2
14	Iceland	76.5	34	Norway	69.0
15	Luxembourg	75.2	35	Slovakia	68.7
16	Finland	74.8	36	Botswana	68.6
17	Japan	72.5	37	Czech Republic	68.5
18	Mauritius	72.3	38	Kuwait	68.3
19	Bahrain	72.2		Latvia	68.3
20	Belgium	71.5	40	Uruguay	68.1

Privacy[b]

Highest, 2007 (1=highest surveillance; 5=lowest surveillance)

1	China	1.3		Sweden	2.1
	Malaysia	1.3	20	Australia	2.2
	Russia	1.3		Israel	2.2
4	Singapore	1.4		Japan	2.2
	United Kingdom	1.4		Latvia	2.2
6	Taiwan	1.5	24	Austria	2.3
	Thailand	1.5		Cyprus	2.3
	United States	1.5		New Zealand	2.3
9	Philippines	1.8		Poland	2.3
10	France	1.9		South Africa	2.3
	India	1.9		Spain	2.3
12	Bulgaria	2.0	30	Malta	2.4
	Denmark	2.0		Switzerland	2.4
	Lithuania	2.0	32	Czech Republic	2.5
15	Brazil	2.1		Finland	2.5
	Netherlands	2.1		Ireland	2.5
	Norway	2.1	35	Belgium	2.7
	Slovakia	2.1		Iceland	2.7

a Ranks countries on the basis of indicators of how government intervention can restrict the economic relations between individuals, published by the Heritage Foundation. The ranking includes data on labour and business freedom as well as trade policy, taxation, monetary policy, the banking system, foreign-investment rules, property rights, the amount of economic output consumed by the government, regulation policy, the size of the black market and the extent of wage and price controls. Countries are scored from 80–100 (free) to 0–49.9 (repressed).

b Ranks EU countries and a benchmark group of non-EU countres. Levels of surveillance are ranked by combining criteria such as: level of legal protection; privacy enforcement; use of identity cards and biometrics; data-sharing; visual surveillance; communication interception; workplace monitoring.

Economic growth

Highest economic growth, 1996–2006
Average annual % increase in real GDP

1	Equatorial Guinea	33.7	28	Bahrain	7.2
2	Liberia	15.0		Lithuania	7.2
3	Azerbaijan	13.4	30	Georgia	7.1
4	Turkmenistan	13.3		Laos	7.1
5	Myanmar	12.6	32	Belize	6.8
6	Qatar[a]	11.7	33	Burkina Faso	6.6
7	Armenia	10.5		Dominican Republic	6.6
8	China	10.3	35	Tanzania	6.4
9	Cambodia	9.9	36	Kuwait[a]	6.3
10	Trinidad & Tobago	9.6		Uganda	6.3
11	Mozambique	9.4	38	Bangladesh	6.2
12	Chad	9.3	39	Luxembourg	6.0
13	Angola	9.0	40	Costa Rica	5.9
14	Estonia	8.7		Mongolia	5.9
15	Bhutan	8.6		Singapore	5.9
	Latvia	8.6	43	Albania	5.7
17	Kazakhstan	8.2		Gambia, The	5.7
18	Tajikistan	8.1		Jordan	5.7
19	Ireland	8.0	46	Egypt	5.6
	Nigeria	8.0		Mali	5.6
21	Belarus	7.9		Panama	5.6
	Vietnam	7.9	49	Ethiopia	5.5
23	Rwanda	7.7		Russia	5.5
24	Botswana	7.5		Uzbekistan	5.5
25	United Arab Emirates	7.4	52	Ghana	5.4
26	India	7.3		Sri Lanka	5.4
	Sudan	7.3		Tunisia	5.4

Lowest economic growth, 1996–2006
Average annual % change in real GDP

1	Congo-Kinshasa[a]	0.8	20	Austria	2.5
	Côte d'Ivoire	0.8		Portugal	2.5
	Gabon	0.8	22	Barbados	2.6
4	Eritrea	0.9		Belgium	2.6
	Haiti	0.9		France	2.6
	Papua New Guinea	0.9	25	Syria[a]	2.7
7	Jamaica	1.0	26	Brazil	2.8
8	Japan	1.2		Indonesia	2.8
9	Togo	1.4		Romania	2.8
10	Germany	1.6		Swaziland	2.8
11	Central African Rep	1.7	30	Argentina	2.9
	Italy	1.7		Colombia	2.9
	Paraguay	1.7		Netherlands	2.9
14	Uruguay	1.8	33	Lesotho	3.0
15	Burundi[a]	2.1		Norway	3.0
	Switzerland	2.1		Thailand	3.0
	West Bank and Gaza	2.1		Venezuela	3.0
18	Denmark	2.4	37	Lebanon[a]	3.1
	Fiji	2.4			

a Estimate.

Highest economic growth, 1986–96
Average annual % increase in real GDP

1	Equatorial Guinea	15.0	11	Botswana	8.5
2	China	11.4	12	Cambodia	8.1
3	Thailand	10.6	13	Vietnam	7.9
4	Singapore	10.3	14	Indonesia	7.7
5	Malaysia	10.1	15	Mauritius	7.4
6	Belize	10.0	16	Uganda	7.2
7	South Korea	9.3	17	Mozambique	6.9
8	Bhutan	8.8		Swaziland	6.9
	Chile	8.8		Syria	6.9
10	Taiwan	8.6	20	United Arab Emirates	6.7

Lowest economic growth, 1986–96
Average annual % change in real GDP

1	Liberia	-21.8	11	Albania	-0.9
2	Bulgaria	-5.3	12	Hungary	-0.8
3	Congo-Kinshasa	-4.7	13	Haiti	-0.7
	Sierra Leone	-4.7	14	Burundi	-0.6
5	Rwanda	-4.3	15	Zambia	-0.3
6	Lebanon	-2.8	16	Mongolia	-0.2
7	Cameroon	-2.5	17	Nicaragua	0.1
8	Romania	-2.0	18	Suriname	0.3
9	Central African Rep	-1.9	19	Czech Republic	0.5
10	Libya	-1.0			

Highest services growth, 1996–2006
Average annual % increase in real terms

1	Azerbaijan	13.7	10	Estonia	8.8
2	Armenia	11.2	11	Mauritania	8.7
	Bhutan	11.2	12	Rwanda	8.6
4	China	11.0	13	Belarus	8.4
5	Cambodia	10.2		Uganda	8.4
6	India	9.6	15	Ireland[a]	7.9
7	Georgia	9.2		Kazakhstan	7.9
8	Latvia	9.1	17	Albania	7.7
9	United Arab Emirates[a]	8.9	18	Dominican Republic	7.6

Lowest services growth, 1996–2006
Average annual % change in real terms

1	Zimbabwe[a]	-5.5	8	Italy	2.0
2	Guinea-Bissau	-1.1		Uruguay	2.0
3	Paraguay	1.4		Thailand	2.0
4	Japan[a]	1.7	11	Austria	2.3
5	Jamaica	1.8	12	Denmark	2.4
6	Germany	1.9	13	Belgium	2.6
	Switzerland[a]	1.9		Guinea	2.6

a 1995–2005
Note: Rankings of highest and lowest industrial growth 1996–2006 can be found on page 46 and highest and lowest agricultural growth 1996–2006 on page 49.

Trading places

Biggest exporters

% of total world exports (goods, services and income), 2006

1	Euro area (12)	16.60	23	India	1.19
2	United States	12.11	24	Malaysia	1.10
3	Germany	8.90	25	Norway	1.08
4	United Kingdom	6.50	26	Australia	1.04
5	China	6.43	27	Hong Kong	1.03
6	Japan	5.19	28	Denmark	0.99
7	France	4.54	29	Luxembourg	0.97
8	Netherlands	3.47	30	Brazil	0.95
9	Italy	3.40	31	Thailand	0.91
10	Canada	2.98	32	Poland	0.82
11	Belgium	2.37	33	United Arab Emirates	0.81
12	South Korea	2.30	34	Turkey	0.70
13	Spain	2.15	35	Czech Republic	0.66
14	Russia	2.11	36	Finland	0.65
15	Switzerland	1.90	37	Indonesia	0.59
16	Taiwan	1.62	38	Hungary	0.54
17	Mexico	1.58	39	South Africa	0.47
18	Ireland	1.44	40	Kuwait	0.46
19	Sweden	1.41	41	Iran	0.43
20	Singapore	1.34		Portugal	0.43
21	Saudi Arabia	1.32		Venezuela	0.43
22	Austria	1.20	44	Israel	0.41

Most trade dependent

Trade as % of GDP[a], 2006

1	Aruba	172.5
2	Singapore	95.8
3	Malaysia	94.6
4	United Arab Emirates	91.4
5	Slovakia	77.6
6	Swaziland	74.3
7	Belgium	70.9
8	Vietnam	68.9
9	Lesotho	68.7
10	Puerto Rico	68.6
11	Estonia	67.1
12	Hungary	66.3
	Netherlands Antilles	66.3
14	Czech Republic	65.5
15	Equatorial Guinea	64.8
16	Bahrain	63.2
17	Iraq	61.9
18	Tajikistan	61.7
19	Zimbabwe	61.4
20	Taiwan	59.5
21	Slovenia	59.3

Least trade dependent

Trade as % of GDP[a], 2006

1	North Korea	6.2
2	Brazil	10.7
3	United States	11.0
4	Central African Rep	12.1
5	Bermuda	12.4
6	Rwanda	12.7
7	Japan	13.2
8	Greece	13.8
9	Cuba	15.1
10	Burkina Faso	15.3
11	Hong Kong	15.5
12	India	15.9
13	Benin	16.2
14	Colombia	16.3
	Euro area (12)	16.3
16	Australia	16.6
17	Sudan	17.0
18	Burundi	17.1
19	Pakistan	17.2
20	Uganda	17.2
21	Cameroon	17.9

Notes: The figures are drawn wherever possible from balance of payment statistics so have differing definitions from trade statistics taken from customs or similar sources. For Hong Kong and Singapore, domestic exports and retained imports only are used.
a Average of imports plus exports of goods as % of GDP.

Biggest traders of goods
% of world exports of goods, 2006

1	Euro area (12)	15.13	25	Thailand	1.10
2	Germany	9.76	26	Australia	1.08
3	United States	8.86	27	India	1.07
4	China	8.37	28	Norway	1.06
5	Japan	5.31	29	Poland	1.01
6	France	4.15	30	Ireland	0.90
7	United Kingdom	3.88	31	Czech Republic	0.82
8	Italy	3.60	32	Turkey	0.79
9	Canada	3.47	33	Denmark	0.78
10	Netherlands	3.34	34	Indonesia	0.74
11	South Korea	2.86	35	Finland	0.67
12	Russia	2.62	36	Hungary	0.64
13	Belgium	2.43		Iran	0.64
14	Mexico	2.16	38	Venezuela	0.56
15	Taiwan	2.02	39	South Africa	0.55
16	Spain	1.87	40	Puerto Rico	0.52
17	Saudi Arabia	1.82	41	Kuwait	0.51
18	Switzerland	1.44	42	Chile	0.50
19	Malaysia	1.39	43	Algeria	0.47
20	Sweden	1.28	44	Nigeria	0.45
21	Singapore	1.23	45	Argentina	0.40
22	United Arab Emirates	1.20		Philippines	0.40
23	Brazil	1.19	47	Israel	0.38
24	Austria	1.16		Portugal	0.38

Biggest traders of services and income
% of world exports of services and income, 2006

1	Euro area (12)	19.56	24	Russia	1.06
2	United States	18.69	25	Australia	0.96
3	United Kingdom	11.80	26	Taiwan	0.82
4	Germany	7.15	27	Greece	0.70
5	France	5.32	28	Finland	0.60
6	Japan	4.95	29	Malaysia	0.53
7	Netherlands	3.73		Portugal	0.53
8	Italy	2.99	31	Turkey	0.51
9	Switzerland	2.84	32	Thailand	0.50
10	Spain	2.72	33	Israel	0.48
11	Hong Kong	2.71	34	Brazil	0.45
12	Luxembourg	2.64	35	Poland	0.43
13	Ireland	2.53	36	Mexico	0.40
14	China	2.50	37	Kuwait	0.38
15	Belgium	2.27	38	Czech Republic	0.33
16	Canada	1.99		Egypt	0.33
17	Sweden	1.67		Hungary	0.33
18	Singapore	1.56	41	Saudi Arabia	0.31
19	India	1.45		South Africa	0.31
20	Denmark	1.40	43	Indonesia	0.27
21	Austria	1.27	44	Lebanon	0.24
22	South Korea	1.14	45	Argentina	0.23
23	Norway	1.12	46	Ukraine	0.22

Balance of payments: current account

Largest surpluses
$m, 2006

1	China	249,866	26	Belgium	10,671
2	Japan	170,520	27	Austria	10,259
3	Germany	150,750	28	Indonesia	9,937
4	Saudi Arabia	99,066	29	Argentina	8,092
5	Russia	94,257	30	Israel	7,990
6	Norway	58,323	31	Denmark	7,339
7	Netherlands	55,795	32	South Korea	6,092
8	Switzerland	54,849	33	Philippines	5,897
9	Kuwait	50,996	34	Iraq[a]	5,666
10	Singapore	36,326	35	Chile	5,256
11	United Arab Emirates	35,942	36	Brunei	5,232
12	Algeria	28,950	37	Trinidad & Tobago	4,654
13	Sweden	28,413	38	Oman	4,377
14	Venezuela	27,149	39	Luxembourg	4,370
15	Malaysia	25,488	40	Azerbaijan	3,708
16	Taiwan	24,655	41	Turkmenistan	3,351
17	Libya	22,170	42	Uzbekistan	3,198
18	Canada	20,797	43	Macau	2,946
19	Iran[a]	20,650	44	Egypt	2,635
20	Hong Kong	20,151	45	Peru	2,589
21	Qatar	16,113	46	Botswana	1,940
22	Nigeria[a]	13,800	47	Bahrain	1,918
23	Brazil	13,620	48	Morocco	1,778
24	Finland	10,878	49	Ecuador	1,503
25	Angola	10,690	50	Bolivia	1,319

Largest deficits
$m, 2006

1	United States	-811,490	23	Iceland	-4,234
2	Spain	-106,344	24	Serbia	-3,966
3	United Kingdom	-77,550	25	Slovakia	-3,919
4	Italy	-47,312	26	Croatia	-3,220
5	Australia	-41,240	27	Lithuania	-3,218
6	Turkey	-32,774	28	Colombia	-3,057
7	Greece	-29,565	29	Estonia	-2,446
8	France	-28,310	30	Thailand	-2,175
9	Portugal	-18,281	31	Mexico	-2,008
10	South Africa	-16,487	32	Jordan	-1,909
11	Romania	-12,785	33	Kazakhstan	-1,795
12	Euro area (12)	-12,490	34	Ethiopia	-1,786
13	Poland	-11,084	35	Ukraine	-1,617
14	India	-9,415	36	Guatemala	-1,592
15	New Zealand	-9,381	37	Bahamas	-1,567
16	Ireland	-9,095	38	Belarus	-1,512
17	Hungary	-7,421	39	Lebanon	-1,451
18	Pakistan	-6,795	40	Tanzania	-1,442
19	Sudan	-5,110	41	Sri Lanka	-1,434
20	Bulgaria	-5,010	42	Georgia	-1,236
21	Czech Republic	-4,586	43	Bosnia	-1,233
22	Latvia	-4,522	44	Jamaica	-1,170

Largest surpluses as % of GDP
%, 2006

1	Kuwait	49.9	26	Hong Kong	10.6
2	Brunei	45.3	27	Luxembourg	10.5
3	Libya	44.1	28	Russia	9.6
4	Turkmenistan	31.9	29	Aruba	9.5
5	Qatar	30.6		Iran[a]	9.5
6	Saudi Arabia	28.4	31	China	9.4
7	United Arab Emirates	27.7	32	Netherlands	8.4
8	Singapore	27.5	33	Myanmar	8.0
9	Trinidad & Tobago	25.7	34	Sweden	7.4
10	Algeria	25.2	35	Zambia	6.9
11	Angola	23.7	36	Taiwan	6.8
12	Macau	20.8	37	Israel	5.7
13	Azerbaijan	18.7	38	Finland	5.2
14	Uzbekistan	18.6		Germany	5.2
15	Botswana	18.3		Suriname	5.2
16	Norway	17.4	41	Philippines	5.0
17	Malaysia	16.9	42	Equatorial Guinea	4.5
18	Namibia	15.2		Lesotho	4.5
19	Venezuela	14.9	44	Japan	3.9
20	Switzerland	14.4	45	Argentina	3.8
21	Oman	14.2	46	Swaziland	3.7
22	Iraq[a]	13.4	47	Chile	3.6
23	Bahrain	12.0		Ecuador	3.6
	Nigeria	12.0	49	Mongolia	3.5
25	Bolivia	11.8	50	Austria	3.2

Largest deficits as % of GDP
%, 2006

1	Liberia	-58.4		Moldova	-11.7
2	Burundi	-36.0	22	Tanzania	-11.3
3	West Bank and Gaza[b]	-27.3	23	Lithuania	-10.8
4	Iceland	-26.0	24	Romania	-10.5
5	Bahamas	-25.4	25	Burkina Faso[b]	-10.3
6	Fiji	-22.8	26	Bosnia	-10.1
7	Latvia	-22.5		Sierra Leone	-10.1
8	Nicaragua	-16.1	28	Senegal	-9.7
9	Georgia	-16.0	29	Greece	-9.6
10	Bulgaria	-15.9		Mauritius	-9.6
11	Laos[b]	-15.8	31	Portugal	-9.4
12	Estonia	-14.9	32	Mozambique	-9.3
13	Zimbabwe[a]	-14.5	33	Chad	-9.2
14	Kyrgyzstan	-14.3	34	New Zealand	-9.0
15	Gambia, The	-14.2	35	Madagascar	-8.7
16	Sudan	-13.6		Spain	-8.7
17	Jordan	-13.5	37	Niger	-8.4
18	Ethiopia	-13.4	38	Barbados	-8.1
19	Serbia	-12.4		Ghana	-8.1
20	Jamaica	-11.7		Turkey	-8.1

a Estimate.   b 2005

Workers' remittances
$m, 2006

1	India	25,426	16	Pakistan	5,121
2	Mexico	24,732	17	Vietnam	4,800
3	China	23,319	18	Serbia and Mont.	4,703
4	Philippines	15,250	19	Poland	4,370
5	France	12,479	20	Brazil	4,253
6	Spain	8,863	21	Colombia	3,929
7	Belgium	7,220	22	Guatemala	3,626
8	United Kingdom	6,954	23	El Salvador	3,330
9	Romania	6,707	24	Nigeria	3,329
10	Germany	6,667	25	Portugal	3,329
11	Indonesia	5,722	26	Australia	3,119
12	Morocco	5,454	27	Russia	3,091
13	Bangladesh	5,428	28	Austria	3,084
14	Egypt	5,330	29	Dominican Republic	3,044
15	Lebanon	5,183	30	Ecuador	2,922

Official reserves[a]
$m, end-2006

1	China	1,080,762	16	Italy	75,774
2	Japan	895,320	17	Thailand	67,007
3	Russia	303,773	18	Switzerland	64,463
4	Taiwan	266,864	19	Turkey	63,265
5	South Korea	239,148	20	Libya	62,227
6	United States	221,086	21	Norway	56,842
7	India	178,050	22	Australia	55,082
8	Singapore	136,259	23	Poland	48,474
9	Hong Kong	133,213	24	United Kingdom	47,038
10	Germany	111,639	25	Nigeria	42,736
11	France	98,238	26	Indonesia	42,597
12	Brazil	85,843	27	Venezuela	36,715
13	Malaysia	82,876	28	Canada	35,064
14	Algeria	81,461	29	Argentina	32,022
15	Mexico	76,330	30	Czech Republic	31,457

Gold reserves
Market value, $m, end-2006

1	United States	166,236	14	United Kingdom	6,338
2	Germany	69,952	15	Austria	5,899
3	France	55,586	16	Lebanon	5,861
4	Italy	50,112	17	Belgium	4,653
5	Switzerland	26,369	18	Algeria	3,547
6	Japan	15,638	19	Sweden	3,240
7	Netherlands	13,102	20	Philippines	2,938
8	China	12,269	21	Libya	2,937
9	Spain	8,518	22	Saudi Arabia	2,922
10	Russia	8,206	23	South Africa	2,536
11	Portugal	7,819	24	Turkey	2,373
12	India	7,312	25	Greece	2,284
13	Venezuela	7,298	26	Romania	2,140

a Foreign exchange, SDRs, IMF position and gold at market prices.

Exchange rates

The Economist's Big Mac index

		Big Mac prices in local currency	in $	Implied PPP[a] of the $	Actual $ exchange rate	Under (−)/ over (+) valuation against $, %
Countries with the most under-valued currencies, July 2007						
1	China	11.00	1.45	3.23	7.60	−58
2	Hong Kong	12.00	1.54	3.52	7.82	−55
3	Malaysia	5.50	1.60	1.61	3.43	−53
4	Egypt	9.54	1.68	2.80	5.69	−51
5	Indonesia	15900.00	1.76	4662.76	9015.00	−48
6	Thailand	62.00	1.80	18.18	34.46	−47
7	Philippines	85.00	1.85	24.93	45.90	−46
	Ukraine	9.25	1.84	2.71	5.03	−46
9	Sri Lanka	210.00	1.89	61.58	111.40	−45
10	Russia	52.00	2.03	15.25	25.65	−41
11	Paraguay	10500.00	2.04	3079.18	5145.00	−40
12	Costa Rica	1130.00	2.18	331.38	518.63	−36
13	South Africa	15.50	2.22	4.55	6.97	−35
14	Japan	280.00	2.29	82.11	122.32	−33
	Taiwan	75.00	2.29	21.99	32.77	−33
16	Pakistan	140.00	2.32	41.06	60.37	−32
17	Saudi Arabia	9.00	2.40	2.64	3.75	−30
18	Czech Republic	52.90	2.51	15.51	21.12	−27
	Slovakia	61.30	2.49	17.98	24.59	−27
19	Poland	6.90	2.51	2.02	2.75	−26

Countries with the most over-valued or least under-valued currencies, July 2007						
1	Iceland	469.00	7.61	137.54	61.67	123
2	Norway	40.00	6.88	11.73	5.81	102
3	Switzerland	6.30	5.20	1.85	1.21	53
4	Denmark	27.75	5.08	8.14	5.46	49
5	Sweden	33.00	4.86	9.68	6.79	42
6	Euro area[b]	3.06	4.17	1.12[c]	1.36[c]	22
7	United Kingdom	1.99	4.01	1.71[d]	2.01[d]	18
8	Canada	3.88	3.68	1.14	1.05	8
9	Turkey	4.75	3.66	1.39	1.30	7
10	Brazil	6.90	3.61	2.02	1.91	6
11	New Zealand	4.60	3.59	1.35	1.28	5
12	Colombia	6900.00	3.53	2023.46	1956.00	3
13	Venezuela	7400.00	3.45	2170.09	2147.30	1
14	Hungary	600.00	3.33	175.95	180.21	−2
15	South Korea	2900.00	3.14	850.44	922.50	−8
16	Peru	9.50	3.00	2.79	3.17	−12
17	Chile	1565.00	2.97	458.94	526.75	−13
18	Australia	3.45	2.95	1.01	1.17	−14

a Purchasing-power parity: local price divided by United States price ($3.41, average of four cities)
b Weighted average of price in euro area
c Dollars per euro
d Dollars per pound

Inflation

Consumer price inflation

Highest, 2007, %		Lowest, 2007, %	
1 Zimbabwe	24,411.0	1 Burkina Faso	-0.2
2 Congo-Kinshasa[a]	21.3	2 Japan	0.1
3 Yemen[b]	20.8	Niger	0.1
4 Myanmar[b]	20.0	4 Israel	0.5
5 Venezuela	18.7	5 Norway	0.7
6 Sri Lanka	17.5	Switzerland	0.7
7 Ethiopia	17.2	7 Cameroon	0.9
Iran	17.2	8 Togo	1.0
9 Azerbaijan	16.7	9 Benin	1.3
10 United Arab Emirates	14.0	Malta	1.3
11 Qatar	13.8	11 Mali	1.4
12 Ukraine	12.8	12 France	1.5
13 Moldova	12.4	13 Netherlands	1.6
14 Guyana	12.3	14 Denmark	1.7
15 Angola	12.2	15 Belgium	1.8
16 Serbia[b]	11.7	Italy	1.8
Sierra Leone	11.7	Peru	1.8
18 Suriname[b]	11.3	18 Côte d'Ivoire	1.9
19 Nicaragua	11.1	Vanuatu[b]	1.9
20 Ghana[b]	10.9	20 Bahrain[b]	2.0
21 Kazakhstan	10.8	Hong Kong	2.0
22 Zambia	10.7	Malaysia	2.0
		Morocco	2.0

Consumer price inflation, 2002–07

Highest average annual consumer price inflation, %		Lowest average annual consumer price inflation, %	
1 Zimbabwe	1,091.2	1 Japan	0.0
2 Angola	34.8	Montenegro	0.0
3 Venezuela	20.1	3 Libya[d]	0.2
4 Haiti	19.4	4 Hong Kong	0.4
5 Dominican Republic	18.1	5 Israel	0.8
6 Myanmar[d]	17.0	6 Mali	0.9
7 Ghana[d]	16.2	Switzerland	0.9
8 Zambia	15.4	8 Maldives	1.1
9 Iran	14.7	Singapore	1.1
10 Belarus	14.2	Taiwan	1.1
11 Yemen[d]	13.9	11 Finland	1.2
12 Suriname[d]	13.4	12 Niger	1.3
13 Turkey	12.9	Sweden	1.3
14 Congo-Kinshasa[c]	12.5	14 Guinea Bissau	1.4
15 Ethiopia	12.3	15 Netherlands	1.5
Moldova	12.3	Norway	1.5
17 Serbia[d]	12.2	Oman[d]	1.5
18 Nigeria	12.0	18 Saudi Arabia	1.6
19 Malawi	11.6	19 Cameroon	1.7
20 Jamaica	11.4	Denmark	1.7
		Gabon[d]	1.7
		Germany	1.7

a 2005 b 2006 c 2002–05 d 2002–06

Commodity prices

2007, % change on a year earlier		*2000–07, % change*	
1 Lead	100.8	1 Lead	467.2
2 Tin	65.6	2 Nickel	331.6
3 Palm oil	64.5	3 Copper	293.1
4 Wheat	57.2	4 Zinc	188.8
5 Nickel	53.9	5 Tin	166.7
6 Coconut oil	51.7	6 Palm oil	154.0
7 Soyabean	45.0	7 Gold	149.0
8 Soya oil	44.5	8 Wheat	145.6
9 Corn	43.2	9 Oil[a]	136.3
10 Soya meal	35.1	10 Soya oil	126.2
11 Wool (Aus)	26.4	11 Cocoa	119.5
12 Cocoa	22.7	12 Rubber	112.6
13 Gold	15.1	13 Coconut oil	100.5
14 Coffee	12.5	14 Corn	78.2
15 Cotton	10.2	15 Soyabeans	72.1
16 Oil[a]	8.6	16 Aluminium	70.3
17 Rice	7.6	17 Rice	60.1
18 Copper	5.8	18 Coffee	47.8
19 Hides	4.8	19 Lamb	43.0
20 Beef (US)	3.4	20 Beef (Aus)	41.2
21 Aluminium	2.8	21 Soya meal	40.0
22 Lamb	2.1	22 Wool (Aus)	35.4
23 Beef (Aus)	1.8	23 Beef (US)	29.8
24 Zinc	-0.3	24 Sugar	23.2
25 Rubber	-3.0	25 Hides	12.7
26 Wool (NZ)	-6.5	26 Cotton	7.4

The Economist's house-price indicators

Q4 2007[b], % change on a year earlier		*1997–2007, % change*	
1 Singapore	31.2	1 South Africa	395
2 Hong Kong	24.3	2 Ireland	227
3 Australia	12.3	3 United Kingdom	210
4 Belgium	11.4	4 Spain	190
5 Sweden	11.3	5 Australia	168
6 China	10.5	6 France	150
7 South Africa	9.1	7 Sweden	149
8 New Zealand	8.9	8 Belgium	142
9 Canada	6.0	9 Denmark	129
10 France	5.6	10 New Zealand	124
11 Italy	5.1	11 United States	104
12 Spain	4.8	12 Italy	102
13 United Kingdom	4.2	Netherlands	102
14 Netherlands	3.8	14 Canada	80
15 Denmark	2.1	15 Switzerland	19
16 Switzerland	2.0	16 Hong Kong	-29
17 Japan	-0.7	17 Japan	-32
18 Germany	-2.7		
19 Ireland	-7.3		
20 United States	-8.9		

a West Texas Intermediate. b Or latest.

Debt

Highest foreign debt[a]
$bn, 2006

1	China	322.8	26	Pakistan	35.9
2	Russia	251.1	27	South Africa	35.6
3	Turkey	207.9	28	Egypt	29.3
4	Brazil	194.2	29	Peru	28.2
5	South Korea	187.0	30	Slovakia	27.1
6	Mexico	160.7	31	Kuwait	25.0
7	India	153.1	32	Lebanon	24.0
8	Indonesia	131.0		Singapore	24.0
9	Poland	125.8	34	Latvia	22.8
10	Argentina	122.2	35	Slovenia	21.0
11	Hungary	107.7	36	Bulgaria	20.9
12	Taiwan	91.0	37	Bangladesh	20.5
13	Israel	87.0	38	Vietnam	20.2
14	Kazakhstan	74.1	39	Iran	20.1
15	Philippines	60.3	40	Sudan	19.2
16	Czech Republic	58.0	41	Lithuania	19.0
17	Thailand	55.2	42	Morocco	18.5
18	Romania	55.1		Tunisia	18.5
19	Malaysia	52.5	44	Cuba	17.0
20	Ukraine	49.9	45	Ecuador	16.5
21	Chile	48.0	46	Côte d'Ivoire	13.8
22	United Arab Emirates	46.0		Serbia	13.8
23	Venezuela	44.6	48	Sri Lanka	11.5
24	Colombia	39.7	49	Panama	10.0
25	Croatia	37.5	50	Uruguay	9.8

Highest foreign debt
As % of exports of goods and services, 2006

1	Liberia	2,030	20	Brazil	158
2	Burundi	1,061	21	Cuba	157
3	Eritrea	742	22	Togo	154
4	Central African Rep	597	23	Côte d'Ivoire	150
5	Congo-Kinshasa	388	24	Romania	148
6	Guinea-Bissau	360	25	Jamaica	144
7	Sierra Leone	349	26	Colombia	143
8	Sudan	304	27	Peru	140
9	Latvia	266	28	Nicaragua	131
10	Guinea	261	29	Ecuador	129
11	Zimbabwe	248	30	Lebanon	128
12	Laos	245	31	Hungary	127
13	Argentina	230		Serbia	127
14	Kazakhstan	222	33	Kyrgyzstan	126
15	Myanmar	202	34	Bolivia	123
16	Turkey	200		Pakistan	123
17	Gambia, The	191	36	Indonesia	122
18	Uruguay	185	37	Lithuania	121
19	Croatia	168		Mauritania	121

a Foreign debt is debt owed to non-residents and repayable in foreign currency; the
figures shown include liabilities of government, public and private sectors. Developed
countries have been excluded.

Highest foreign debt burden
Foreign debt as % of GDP, 2006

1	Liberia	674		Kyrgyzstan	71
2	Guinea-Bissau	169	24	Myanmar	70
3	Latvia	135	25	Sudan	69
4	Kazakhstan	132	26	Argentina	68
5	Congo-Kinshasa	130	27	Tunisia	66
6	Burundi	119		Uruguay	66
7	Lebanon	116	29	Moldova	65
8	Zimbabwe	110	30	Gabon	64
9	Gambia	108	31	Israel	61
10	Congo-Brazzaville	104		Turkey	61
11	Hungary	100	33	Mauritania	60
12	Jamaica	99	34	Jordan	58
13	Croatia	93		Romania	58
14	Laos	87		Slovakia	58
15	Sierra Leone	83		Ukraine	58
16	Côte d'Ivoire	82	38	Central African Rep	57
17	Lithuania	79		Philippines	57
18	Panama	77	40	Slovenia	56
19	Bulgaria	74	41	El Salvador	55
	Togo	74	42	Ecuador	52
21	Nicaragua	72		Eritrea	52
22	Guinea	71		Serbia	52

Highest debt service ratios[b]
%, average, 2006

1	Uruguay	87.8	23	Macedonia	15.7
2	Burundi	40.4	24	Tunisia	14.4
3	Brazil	37.3	25	Russia	13.8
4	Kazakhstan	33.7	26	Venezuela	13.3
5	Latvia	33.3	27	El Salvador	13.1
6	Turkey	33.2	28	Peru	12.9
7	Croatia	33.1	29	Angola	12.8
	Hungary	33.1	30	Bulgaria	12.4
9	Argentina	31.6		Gambia	12.4
10	Colombia	31.3	32	Israel	12.2
11	Panama	24.7		Moldova	12.2
	Poland	24.7		Morocco	12.2
13	Ecuador	24.1	35	Jamaica	11.9
14	Lithuania	22.1	36	Czech Republic	10.5
15	Lebanon	21.0	37	Cuba	10.3
16	China	20.0	38	Dominican Republic	9.6
17	Philippines	19.6		Rwanda	9.6
18	Mexico	18.9		Sierra Leone	9.6
	Slovenia	18.9	41	Thailand	9.4
20	Romania	18.4	42	Botswana	8.7
21	Ukraine	18.1	43	Pakistan	8.6
22	Indonesia	16.6		Sri Lanka	8.6

b Debt service is the sum of interest and principal repayments (amortisation) due on
outstanding foreign debt. The debt service ratio is debt service expressed as a
percentage of the country's exports of goods and services.

Aid

Largest bilateral and multilateral donors[a]
$m, 2006

1	United States	23,532	14	Saudi Arabia	2,095
2	United Kingdom	12,459	15	Belgium	1,978
3	Japan	11,187	16	Switzerland	1,646
4	France	10,601	17	Austria	1,498
5	Germany	10,435	18	Ireland	1,022
6	Netherlands	5,452	19	Finland	834
7	Sweden	3,955	20	Turkey	714
8	Spain	3,814	21	Taiwan	513
9	Canada	3,684	22	South Korea	455
10	Italy	3,641	23	Greece	424
11	Norway	2,954	24	Portugal	396
12	Denmark	2,236	25	Poland	297
13	Australia	2,123	26	Luxembourg	291

Largest recipients of bilateral and multilateral aid
$m, 2006

1	Nigeria	11,434	35	Bolivia	581
2	Iraq	8,661		Haiti	581
3	Afghanistan	3,000	37	Jordan	580
4	Pakistan	2,147	38	Turkey	570
5	Sudan	2,058	39	Philippines	562
6	Congo-Kinshasa	2,056	40	Cambodia	529
7	Ethiopia	1,947	41	Nepal	514
8	Vietnam	1,846	42	Bosnia	494
9	Tanzania	1,825	43	Guatemala	487
10	Cameroon	1,684	44	Ukraine	484
11	Mozambique	1,611	45	Peru	468
12	Serbia	1,586	46	Tunisia	432
13	Uganda	1,551	47	Burundi	415
14	Zambia	1,425	48	Niger	401
15	Indonesia	1,405	49	Somalia	392
16	India	1,379	50	Benin	375
17	China	1,245	51	Laos	364
18	Bangladesh	1,223		Sierra Leone	364
19	Ghana	1,176	53	Georgia	361
20	Morocco	1,046	54	Albania	321
21	Colombia	988	55	Kyrgyzstan	311
22	Kenya	943	56	Chad	284
23	Egypt	873		Yemen	284
24	Burkina Faso	871	58	Zimbabwe	280
25	Mali	825	59	Papua New Guinea	279
	Senegal	825	60	Liberia	269
27	Sri Lanka	796	61	Congo-Brazzaville	254
28	Madagascar	754	62	Côte d'Ivoire	251
29	Nicaragua	733	63	Mexico	247
30	South Africa	718	64	Malaysia	240
31	Lebanon	707		Tajikistan	240
32	Malawi	699	66	Moldova	228
33	Honduras	587	67	Armenia	213
34	Rwanda	585	68	Timor-Leste	210

Largest bilateral and multilateral donors[a]
% of GDP, 2006

1	Sweden	1.03	14	Germany	0.36
2	Norway	0.88	15	Spain	0.31
3	Netherlands	0.82	16	Canada	0.29
4	Denmark	0.81	17	Australia	0.27
5	Luxembourg	0.70	18	Iceland	0.26
6	Saudi Arabia	0.60		Japan	0.26
7	United Kingdom	0.52	20	New Zealand	0.25
8	Belgium	0.50	21	Italy	0.20
9	Austria	0.47		Portugal	0.20
	France	0.47	23	United Arab Emirates	0.19
11	Ireland	0.46	24	Turkey	0.18
12	Switzerland	0.43		United States	0.18
13	Finland	0.40	26	Kuwait	0.15

Largest recipients of bilateral and multilateral aid
$ per head, 2006

1	West Bank and Gaza	381.3	34	Kyrgyzstan	58.7
2	Cape Verde	328.4	35	Sudan	55.6
3	Iraq	292.6	36	Moldova	54.3
4	Timor-Leste	209.7	37	Burundi	53.2
5	Lebanon	196.5	38	Ghana	52.0
6	Montenegro	138.6	39	Uganda	51.9
7	Suriname	136.9	40	Guinea-Bissau	51.5
8	Nicaragua	130.8	41	Malawi	50.6
9	Bosnia	126.8	42	Equatorial Guinea	49.7
10	Zambia	119.7	43	Papua New Guinea	46.5
11	Albania	103.5	44	Gambia, The	46.3
12	Cameroon	101.5	45	Somalia	46.1
13	Jordan	99.9	46	Croatia	43.5
	Macedonia	99.9	47	Benin	43.1
15	Afghanistan	96.5	48	Bhutan	42.8
16	Nigeria	85.1	49	Tunisia	42.4
17	Georgia	81.9	50	Lesotho	39.9
18	Mozambique	79.8	51	Madagascar	39.5
19	Honduras	79.4	52	Sri Lanka	38.1
20	Liberia	79.0	53	Guatemala	37.8
21	Mongolia	75.0	54	Cambodia	36.7
22	Armenia	71.1	55	Botswana	36.2
23	Senegal	69.3	56	Congo-Kinshasa	34.7
24	Namibia	69.2	57	Swaziland	34.5
25	Haiti	67.6	58	Morocco	32.8
26	Burkina Faso	64.0	59	Central African Rep	32.7
27	Sierra Leone	63.8	60	Chad	28.4
28	Rwanda	63.6	61	Eritrea	28.1
29	Congo-Brazzaville	62.1	62	Niger	27.9
30	Bolivia	61.8	63	Kenya	26.9
	Fiji	61.8	64	Belize	26.4
32	Laos	59.7	65	Ethiopia	24.6
33	Mali	59.4	66	Azerbaijan	24.2

a China also provides aid, but does not disclose amounts.

Industry and services

Largest industrial output
$bn, 2006

1	United States[a]	2,634	23	Switzerland[a]	98
2	Japan[a]	1,355	24	Sweden	97
3	China	1,279	25	Iran	96
4	Germany	783	26	Poland	94
5	United Kingdom	508	27	Thailand	92
6	Italy	439	28	Austria	89
7	France	417	29	Belgium	85
8	Canada	371	30	Malaysia	75
9	Russia	332	31	United Arab Emirates[a]	74
10	Spain	324	32	Argentina	70
11	South Korea	313		South Africa	70
12	South Asia	291	34	Chile	66
13	Brazil	284	35	Ireland[a]	64
14	India	231	36	Denmark	61
15	Saudi Arabia	227	37	Algeria[a]	59
16	Mexico	201		Finland	59
17	Australia[a]	190	39	Greece	58
18	Indonesia	172	40	Nigeria[a]	55
19	Netherlands	144	41	Czech Republic	50
20	Norway	134	42	Colombia	49
21	Turkey	103	43	Israel	48
22	Taiwan	100	44	Singapore	43

Highest growth in industrial output
Average annual % increase in real terms, 1996–2006

1	Chad	22.7	11	Burkina Faso	10.6
2	Azerbaijan	18.7	12	Kazakhstan	10.5
3	Cambodia	18.1	13	Tajikistan	9.8
	Mozambique	16.1	14	Georgia	9.5
5	Sudan	15.8		Tanzania	9.5
6	Armenia	12.8	16	Ireland[b]	9.4
7	Belarus	12.1	17	Estonia	9.2
8	Angola	11.5	18	Uganda	9.1
	China	11.5	19	Bhutan	8.7
10	Vietnam	11.2		Rwanda	8.7

Lowest growth in industrial output
Average annual % change in real terms, 1996–2006

1	Zimbabwe[b]	-6.1		Paraguay	0.6
2	Burundi[b]	-2.7		United Kingdom	0.6
3	Guinea-Bissau	-1.8	13	Israel	0.9
4	Moldova	-1.7		Switzerland[b]	0.9
5	Gabon	-1.6	15	Norway	1.1
6	Italy	0.4	16	Portugal	1.3
7	Japan[b]	0.5	17	Netherlands	1.4
	Taiwan	0.5	18	Uruguay	1.5
9	Jamaica	0.6		Central African Republic	1.7
	Lebanon	0.6			

a 2005   b 1995–2005

Largest manufacturing output
$bn, 2006

1	United States[a]	1663	21	Switzerland[b]	67
2	Japan[a]	954	22	Sweden[b]	60
3	China[a]	751	23	Belgium[a]	56
4	Germany[a]	584	24	Poland	56
5	Italy	299	25	Austria[a]	53
6	United Kingdom[a]	270	26	Malaysia	45
7	Canada	260	27	Argentina	44
8	France	248	28	Ireland[a]	43
9	South Korea	220	29	South Africa	41
10	Brazil	169	30	Finland[a]	39
11	Russia	164	31	Singapore	36
12	Spain[a]	156	32	Czech Republic	35
13	Mexico	136	33	Saudi Arabia	33
14	India	135	34	Denmark[a]	31
15	Indonesia	102	35	Norway	28
16	Taiwan	85	36	Philippines	27
	Turkey	85	37	Romania	26
18	Netherlands[a]	79	38	Iran	25
19	Australia[a]	75	39	Greece[a]	25
20	Thailand	72			

Largest services output
$bn, 2006

1	United States[a]	8762	27	Hong Kong[a]	155
2	Japan[a]	3111	28	South Africa	150
3	Germany	1803	29	Indonesia	146
4	United Kingdom	1582	30	Portugal	121
5	France	1550	31	Finland	119
6	Italy	1179	32	Saudi Arabia	112
7	China	1056	33	Argentina	111
8	Canada	814	34	Ireland[a]	109
9	Spain	732	35	Iran	97
10	Brazil	588	36	Israel	96
11	Mexico	523	37	Thailand	92
12	Russia	470	38	Singapore	81
13	Australia[a]	467	39	New Zealand	76
14	India	452	40	Czech Republic	75
15	South Korea	451	41	Colombia	72
16	Netherlands	430	42	Chile	66
17	Taiwan	266	43	Hungary	64
18	Belgium	262		Pakistan	64
19	Turkey	243		Philippines	64
20	Switzerland[a]	242	46	Malaysia	62
21	Sweden	233	47	United Arab Emirates[a]	56
22	Greece	212	48	Romania	54
23	Austria	195	49	Ukraine	52
24	Poland	191	50	Egypt	48
25	Denmark	169	51	Peru	46
26	Norway	160	52	Venezuela[b]	42

a 2005   b 2004

Agriculture

Largest agricultural output
$bn, 2006

1	China	310	16	Pakistan	23
2	India	145		Nigeria[a]	23
3	United States[a]	137	18	Iran	22
4	Japan[a]	68	19	Thailand	22
5	Brazil	47	20	Australia[a]	21
	Indonesia	47	21	United Kingdom	20
7	France	41	22	Argentina	17
	Russia	41		Colombia	17
9	Turkey	37		Philippines	17
10	Italy	34	25	Egypt	14
	Spain	34		Poland	14
12	Mexico	29	27	Malaysia	13
13	Canada	28		Netherlands	13
14	South Korea	26	29	Bangladesh	12
15	Germany	25		Vietnam	12

Most economically dependent on agriculture
% of GDP from agriculture, 2006

1	Liberia[a]	66.0	14	Afghanistan[a]	36.1
2	Guinea-Bissau	61.8	15	Burundi[a]	34.9
3	Central African Republic	55.8	16	Nepal	34.4
4	Ethiopia	47.3	17	Malawi	34.2
5	Sierra Leone	46.4	18	Kyrgyzstan	33.0
6	Congo-Kinshasa	45.7	19	Gambia, The[a]	32.7
7	Tanzania	45.3	20	Uganda	32.3
8	Togo[a]	43.6	21	Sudan	32.3
9	Laos	42.0	22	Benin[a]	32.2
10	Papua New Guinea[b]	41.8		Timor-Leste	32.2
11	Rwanda	41.0	24	Burkina Faso	30.7
12	Ghana	37.4	25	Cambodia	30.1
13	Mali	36.9	26	Mozambique	28.3

Least economically dependent on agriculture
% of GDP from agriculture, 2006

1	Macau	0.0	14	Japan[a]	1.5
2	Hong Kong[a]	0.1	15	Norway	1.6
	Singapore	0.1	16	Austria	1.7
	Taiwan	0.1		Denmark	1.7
5	Luxembourg	0.4	18	Oman[b]	1.9
6	Trinidad & Tobago	0.6	19	Botswana	2.0
7	Brunei	0.7	20	France	2.1
8	United Kingdom	0.9		Ireland[a]	2.1
9	Belgium	1.0		Italy	2.1
	Germany	1.0	23	Canada	2.3
11	United States[a]	1.2		Netherlands	2.3
12	Switzerland[a]	1.3		Slovenia	2.3
13	Sweden	1.4		United Arab Emirates[a]	2.3

a 2005   b 2004

Highest growth in agriculture
Average annual % increase in real terms, 1996–2006

1	Angola	12.1		Mozambique[a]	6.2
2	United Arab Emirates[a]	9.3	11	United States	6.0
3	Tajikistan	9.0	12	Azerbaijan	5.9
4	Belize	7.6	13	Armenia	5.8
5	Slovakia	7.0	14	Algeria[a]	5.6
6	Uzbekistan	6.4	15	Burkina Faso	5.5
7	Gambia, The	6.3		Nigeria[a]	5.5
	Rwanda	6.3		Chile	5.5
9	Benin[a]	6.2	18	Cameroon	5.4

Lowest growth in agriculture
Average annual % change in real terms, 1996–2006

1	Mauritania	-3.4	9	Congo-Kinshasa	-1.2
2	Jamaica	-2.9		Zimbabwe[a]	-1.2
3	Luxembourg	-2.7	11	Burundi[a]	-1.1
4	Greece	-2.1	12	Fiji	-0.8
5	Botswana	-2.0	13	Ireland[a]	-0.7
6	Portugal	-1.6	14	Singapore	-0.6
7	Trinidad & Tobago	-1.5		Slovenia	-0.6
8	Switzerland[a]	-1.4			

Biggest producers
'000 tonnes, 2006

Cereals

1	China	445,355	6	France	61,813
2	United States	346,562	7	Brazil	59,017
3	India	239,130	8	Canada	50,895
4	Russia	76,866	9	Bangladesh	45,010
5	Indonesia	66,011	10	Germany	43,475

Meat

1	China	81,733	6	Mexico	5,331
2	United States	41,081	7	Spain	5,293
3	Brazil	19,783	8	France	5,206
4	Germany	6,867	9	Russia	5,153
5	India	6,103	10	Argentina	4,537

Fruit

1	China	93,410	6	Spain	16,514
2	India	43,525	7	Indonesia	15,406
3	Brazil	37,736	8	Mexico	15,385
4	United States	27,328	9	Iran	13,848
5	Italy	17,812	10	Philippines	13,582

Vegetables

1	China	448,446	5	Egypt	16,165
2	India	81,947	6	Russia	15,930
3	United States	37,052	7	Iran	15,760
4	Turkey	25,723	8	Italy	15,133

a 1995–2005

Commodities

Wheat

Top 10 producers , 2006–07
'000 tonnes

1	EU27	125,095
2	China	104,470
3	India	69,350
4	United States	49,316
5	Russia	45,006
6	Canada	25,265
7	Pakistan	21,700
8	Turkey	17,500
9	Ukraine	13,809
10	Australia	10,641

Top 10 consumers, 2006–07
'000s tonnes

1	EU27	123,690
2	China	98,460
3	India	74,140
4	Russia	36,370
5	United States	31,000
6	Pakistan	21,660
7	Turkey	16,980
8	Egypt	15,430
9	Iran	15,000
10	Ukraine	11,300

Rice[a]

Top 10 producers, 2006–07
'000 tonnes

1	China	127,800
2	India	93,350
3	Indonesia	33,300
4	Bangladesh	29,000
5	Vietnam	22,894
6	Thailand	18,250
7	Myanmar	10,600
8	Philippines	10,085
9	Japan	7,786
10	Brazil	7,700

Top 10 consumers, 2006–07
'000 tonnes

1	China	127,800
2	India	86,940
3	Indonesia	35,550
4	Bangladesh	29,764
5	Vietnam	18,747
6	Philippines	11,551
7	Myanmar	10,560
8	Thailand	9,870
9	Brazil	8,719
10	Japan	8,250

Sugar[b]

Top 10 producers, 2006
'000 tonnes

1	Brazil	31,622
2	India	22,347
3	EU25	18,098
4	China	10,682
5	United States	7,034
6	Thailand	5,646
7	Mexico	5,412
8	Australia	4,729
9	Russia	3,459
10	Pakistan	3,263

Top 10 consumers, 2006
'000 tonnes

1	India	20,110
2	EU25	17,527
3	China	11,975
4	Brazil	11,262
5	United States	9,228
6	Russia	6,500
7	Mexico	4,979
8	Indonesia	4,195
9	Pakistan	3,951
10	Egypt	2,700

Coarse grains[c]

Top 5 producers, 2006–07
'000 tonnes

1	United States	280,300
2	China	155,400
3	EU27	140,500
4	Brazil	53,700
5	India	33,600

Top 5 consumers, 2006–07
'000 tonnes

1	United States	242,700
2	China	152,700
3	EU27	149,700
4	Brazil	45,200
5	Mexico	40,500

Tea

Top 10 producers, 2006 '000 tonnes		Top 10 consumers, 2006 '000 tonnes	
1 China	1,028	1 India	771
2 India	956	2 China	622
3 Kenya	311	3 Russia	167
Sri Lanka	311	4 Japan	146
5 Turkey	142	5 United Kingdom	135
6 Indonesia	140	6 Turkey	133
7 Vietnam	132	7 Pakistan	117
8 Japan	100	8 United States	108
9 Argentina	80	9 Egypt	79
10 Bangladesh	53	10 Iraq	66

Coffee

Top 10 producers, 2006–07 '000 tonnes		Top 10 consumers, 2006 '000s tonnes	
1 Brazil	2,553	1 United States	1,240
2 Vietnam	1,107	2 Brazil	966
3 Colombia	767	3 Germany	549
4 Indonesia	399	4 Japan	436
5 India	285	5 Italy	336
6 Ethiopia	278	6 France	317
7 Mexico	252	7 Russia	191
8 Guatemala	237	8 Canada	186
9 Honduras	208	9 United Kingdom	183
10 Côte d'Ivoire	171	10 Spain	181

Cocoa

Top 10 producers, 2005–06 '000 tonnes		Top 10 consumers, 2005–06 '000 tonnes	
1 Côte d'Ivoire	1,408	1 United States	822
2 Ghana	741	2 Germany	310
3 Indonesia	560	3 France	239
4 Nigeria	200	4 United Kingdom	225
5 Cameroon	166	5 Russia	178
6 Brazil	162	6 Japan	165
7 Ecuador	114	7 Italy	111
8 Papua New Guinea	51	8 Spain	100
9 Colombia	37	9 Brazil	99
10 Mexico	34	10 Canada	74

a Milled.
b Raw.
c Includes: maize (corn), barley, sorghum, rye, oats and millet.

Copper

Top 10 producers[a], 2006		*Top 10 consumers[b], 2006*	
'000 tonnes		*'000 tonnes*	
1 Chile	5,361	1 China	3,610
2 United States	1,221	2 United States	2,128
3 Peru	1,050	3 Germany	1,394
4 Australia	875	4 Japan	1,282
5 Indonesia	817	5 South Korea	828
6 Russia	779	6 Italy	801
7 China	755	7 Russia	693
8 Canada	603	8 Taiwan	643
9 Zambia	516	9 France	460
10 Poland	497	10 India	407

Lead

Top 10 producers[a], 2006		*Top 10 consumers[b], 2006*	
'000 tonnes		*'000 tonnes*	
1 China	1,490	1 China	2,228
2 Australia	668	2 United States	1,560
3 United States	419	3 Germany	396
4 Peru	313	4 South Korea	325
5 Mexico	134	5 Japan	304
6 Canada	83	6 United Kingdom	300
7 India	69	7 Mexico	292
8 Ireland	62	8 Italy	284
9 Sweden	56	9 Spain	265
10 Morocco	53	10 France	207

Zinc

Top 10 producers[a], 2006		*Top 10 consumers[c], 2006*	
'000 tonnes		*'000 tonnes*	
1 China	2,996	1 China	3,115
2 Australia	1,364	2 United States	1,150
3 Peru	1,202	3 Japan	598
4 United States	727	4 Germany	564
5 Canada	638	5 South Korea	508
6 India	536	6 India	444
7 Mexico	477	7 Belgium	360
8 Kazakhstan	450	8 Italy	313
9 Ireland	426	9 Taiwan	282
10 Sweden	210	10 Australia	280

Tin

Top 5 producers[a], 2006		*Top 5 consumers[b], 2006*	
'000 tonnes		*'000 tonnes*	
1 Indonesia	117.5	1 China	114.8
2 China	114.3	2 United States	47.0
3 Peru	38.5	3 Japan	38.5
4 Bolivia	17.4	4 Germany	20.6
5 Brazil	11.7	5 Taiwan	17.6

Nickel

Top 10 producers[a], 2006		*Top 10 consumers[b], 2006*	
'000 tonnes		'000 tonnes	
1 Russia	290.0	1 China	240.1
2 Canada	232.9	2 Japan	181.3
3 Australia	185.0	3 United States	144.5
4 Indonesia	157.2	4 Taiwan	106.9
5 New Caledonia	103.0	5 Germany	106.2
6 Cuba	74.0	6 South Korea	92.9
7 China	68.9	7 Italy	68.0
8 Philippines	58.9	8 Belgium	58.3
9 Colombia	51.1	9 South Africa	53.5
10 South Africa	41.6	10 Spain	52.5

Aluminium

Top 10 producers[d], 2006		*Top 10 consumers[e], 2006*	
'000 tonnes		'000 tonnes	
1 China	9,349	1 China	8,648
2 Russia	3,718	2 United States	6,150
3 Canada	3,051	3 Japan	2,323
4 United States	2,281	4 Germany	1,823
5 Australia	1,932	5 South Korea	1,153
6 Brazil	1,604	6 India	1,080
7 Norway	1,427	7 Russia	1,047
8 India	1,105	8 Italy	1,021
9 South Africa	887	9 Canada	846
10 Bahrain	844	10 Brazil	773

Precious metals

Gold [a]		*Silver [a]*	
Top 10 producers, 2006		*Top 10 producers, 2006*	
tonnes		tonnes	
1 South Africa	275.1	1 Peru	3,471
2 United States	248.9	2 Mexico	2,999
3 Australia	247.0	3 China	2,000
4 China	226.9	4 Australia	1,727
5 Peru	203.3	5 Chile	1,607
6 Russia	164.3	6 Poland	1,306
7 Canada	104.4	7 United States	1,139
8 Uzbekistan	86.0	8 Canada	995
9 Indonesia	79.6	9 Kazakhstan	811
10 Ghana	69.8	10 Bolivia	473

Platinum		*Palladium*	
Top 3 producers, 2006		*Top 3 producers, 2006*	
tonnes		tonnes	
1 South Africa	164.5	1 Russia	121.3
2 Russia	27.7	2 South Africa	90.4
3 United States/Canada	10.7	3 United States/Canada	30.6

a Mine production. b Refined consumption. c Slab consumption.
d Primary refined production. e Primary refined consumption.

Rubber (natural and synthetic)

Top 10 producers, 2006		*Top 10 consumers, 2006*	
'000 tonnes		*'000 tonnes*	
1 Thailand	3,293	1 China	5,280
2 Indonesia	2,683	2 United States	3,004
3 United States	2,606	3 Japan	2,045
4 China	2,346	4 India	1,079
5 Japan	1,607	5 Germany	904
6 Malaysia	1,335	6 South Korea	727
7 Russia	1,219	7 Brazil	712
8 India	954	8 Russia	613
9 Germany	865	9 France	530
10 France	664	10 Malaysia	496

Raw wool

Top 10 producers[a], 2006		*Top 10 consumers[a], 2006*	
'000 tonnes		*'000 tonnes*	
1 Australia	327	1 China	389
2 China	176	2 India	140
3 New Zealand	168	3 Italy	94
4 Argentina	48	4 Turkey	50
5 India	38	5 Japan	27
6 Uruguay	34	6 Iran	26
7 United Kingdom	30	New Zealand	26
8 South Africa	28	Russia	26
9 Iran	24	United Kingdom	26
10 Russia	21	10 Germany	22

Cotton

Top 10 producers, 2006–07		*Top 10 consumers, 2006–07*	
'000 tonnes		*'000 tonnes*	
1 China	7,966	1 China	10,800
2 India	4,760	2 India	3,932
3 United States	4,700	3 Pakistan	2,660
4 Pakistan	2,091	4 Turkey	1,600
5 Brazil	1,524	5 United States	1,077
6 Uzbekistan	1,171	6 Brazil	900
7 Turkey	820	7 Bangladesh	520
8 Greece	320	8 Indonesia	480
9 Australia	295	9 Mexico	446
10 Burkina Faso	282	10 Thailand	430

Major oil seeds[b]

Top 5 producers, 2006–07		*Top 5 consumers, 2006–07*	
'000 tonnes		*'000 tonnes*	
1 United States	96,250	1 China	86,015
2 Brazil	61,590	2 United States	63,962
3 Argentina	52,855	3 EU25	40,495
4 China	52,820	4 Argentina	38,815
5 India	28,530	5 Brazil	35,876

Oil[c]

Top 10 producers, 2006
'000 barrels per day

1	Saudi Arabia[d]	10,859
2	Russia	9,769
3	United States	6,871
4	Iran[d]	4,343
5	China	3,684
6	Mexico	3,683
7	Canada	3,147
8	United Arab Emirates[d]	2,969
9	Venezuela[d]	2,824
10	Norway	2,778

Top 10 consumers, 2006
'000 barrels per day

1	United States	20,589
2	China	7,445
3	Japan	5,164
4	Russia	2,735
5	Germany	2,622
6	India	2,575
7	South Korea	2,312
8	Canada	2,222
9	Brazil	2,097
10	Saudi Arabia	2,005

Natural gas

Top 10 producers, 2006
Billion cubic metres

1	Russia	612.1
2	United States	524.1
3	Canada	187.0
4	Iran[d]	105.0
5	Norway	87.6
6	Algeria[d]	84.5
7	United Kingdom	80.0
8	Indonesia[d]	74.0
9	Saudi Arabia[d]	73.7
10	Turkmenistan	62.2

Top 10 consumers, 2006
Billion cubic metres

1	United States	619.7
2	Russia	432.1
3	Iran[d]	105.1
4	Canada	96.6
5	United Kingdom	90.8
6	Germany	87.2
7	Japan	84.6
8	Italy	77.1
9	Saudi Arabia[d]	73.7
10	Ukraine	66.4

Coal

Top 10 producers, 2006
Million tonnes oil equivalent

1	China	1,212.3
2	United States	595.1
3	India	209.7
4	Australia	203.1
5	South Africa	144.8
6	Russia	144.5
7	Indonesia[d]	119.9
8	Poland	67.0
9	Germany	50.3
10	Kazakhstan	49.2

Top 10 consumers, 2006
Million tonnes oil equivalent

1	China	1,191.3
2	United States	567.3
3	India	237.7
4	Japan	119.1
5	Russia	112.5
6	South Africa	93.8
7	Germany	82.4
8	Poland	58.4
9	South Korea	54.8
10	Australia	51.1

Oil[c]

Top proved reserves, end 2006
% of world total

1	Saudi Arabia[d]	21.9		5	United Arab Emirates[d]	8.1
2	Iran[d]	11.4		6	Russia	6.6
3	Iraq[d]	9.5			Venezuela[d]	6.6
4	Kuwait[d]	8.4		8	Libya	3.4

a Clean basis.   b Soybeans, sunflower seed, cottonseed, groundnuts and rapeseed.
c Includes crude oil, shale oil, oil sands and natural gas liquids.   d Opec members.

Energy

Largest producers
Million tonnnes oil equivalent, 2005

1	China	1,641	16	Algeria	175
2	United States	1,631	17	United Arab Emirates	168
3	Russia	1,185	18	South Africa	159
4	Saudi Arabia	577	19	Kuwait	146
5	India	419	20	France	137
6	Canada	401	21	Germany	135
7	Iran	304	22	Kazakhstan	122
8	Australia	271	23	Japan	100
9	Indonesia	263	24	Iraq	96
10	Mexico	259	25	Libya	95
11	Norway	234	26	Malaysia	94
12	Nigeria	232	27	Argentina	81
13	Venezuela	205		Ukraine	81
14	United Kingdom	204	29	Poland	79
15	Brazil	188	30	Egypt	76

Largest consumers
Million tonnnes oil equivalent, 2005

1	United States	2,340	16	Spain	145
2	China	1,717	17	Ukraine	143
3	Russia	647	18	Saudi Arabia	140
4	India	537	19	South Africa	128
5	Japan	531	20	Australia	122
6	Germany	345	21	Nigeria	104
7	France	276	22	Thailand	100
8	Canada	272	23	Poland	93
9	United Kingdom	234	24	Turkey	85
10	South Korea	214	25	Netherlands	82
11	Brazil	210	26	Pakistan	76
12	Italy	185	27	Argentina	64
13	Indonesia	180	28	Egypt	61
14	Mexico	177		Malaysia	61
15	Iran	163		Venezuela	61

Energy efficiency[a]

Most efficient
GDP per unit of energy use, 2005

Least efficient
GDP per unit of energy use, 2005

1	Hong Kong	13.5	1	Zimbabwe	0.2
2	Peru	12.7	2	Congo-Kinshasa	0.9
3	Botswana	11.6	3	Uzbekistan	1.1
4	Uruguay	10.6	4	Mozambique	1.4
5	Greece	10.5	5	Trinidad & Tobago	1.6
	Panama	10.5	6	Tanzania	1.8
7	Gabon	10.4		Ukraine	1.8
8	Ireland	10.3	8	Zambia	1.9
9	Costa Rica	9.9	9	Ethiopia	2.0
10	Congo-Brazzaville	9.8	10	Nigeria	2.1
11	Switzerland	9.6	11	Togo	2.3

a 2005 PPP$, per kg of oil equivalent.

Net energy importers

% of commercial energy use, 2005

Highest			Lowest		
1	Hong Kong	100	1	Congo-Brazzaville	-1,041
	Singapore	100	2	Norway	-627
3	Moldova	98	3	Angola	-614
4	Jordan	96	4	Gabon	-604
	Lebanon	96	5	Kuwait	-420
6	Morocco	93	6	Algeria	-404
7	Ireland	89	7	Libya	-399
	Israel	89	8	Oman	-327
9	Jamaica	87	9	Saudi Arabia	-311
	Portugal	87	10	Turkmenistan	-274
11	Belarus	86	11	United Arab Emirates	-258

Largest consumption per head

Kg of oil equivalent, 2005

1	United Arab Emirates	11,436	12	Oman	5,570
2	Kuwait	11,100	13	Belgium	5,407
3	Trinidad & Tobago	9,599	14	Netherlands	5,015
4	Canada	8,417	15	France	4,534
5	United States	7,893	16	Russia	4,517
6	Norway	6,948	17	South Korea	4,426
7	Singapore	6,933	18	Czech Republic	4,417
8	Finland	6,664	19	Germany	4,180
9	Saudi Arabia	6,068	20	Austria	4,174
10	Australia	5,978	21	Japan	4,152
11	Sweden	5,782	22	New Zealand	4,090

Sources of electricity

% of total, 2005

Oil			Gas		
1	Yemen	100.0	1	Turkmenistan	100.0
2	Benin	99.1	2	Trinidad and Tobago	99.5
3	Iraq	98.5	3	Moldova	98.1
4	Cuba	97.5	4	United Arab Emirates	97.9
5	Jamaica	96.6	5	Algeria	96.2

Hydropower			Nuclear power		
1	Paraguay	100.0	1	France	79.1
2	Mozambique	99.8	2	Lithuania	71.1
	Nepal	99.8	3	Slovakia	56.5
4	Congo-Brazzaville	99.7	4	Belgium	55.5
	Congo-Kinshasa	99.7	5	Ukraine	47.7

Coal		
1	Botswana	99.4
2	South Africa	94.1
3	Poland	93.4
4	Estonia	91.2
5	Australia	80.1

Workers of the world[a]

Highest % of population in labour force

1	Cayman Islands	69.2	21	Russia	52.0
2	Laos	66.6	22	Australia	51.8
3	Bhutan	60.4	23	Slovenia	51.3
4	Bermuda	59.4	24	Estonia	51.1
5	Switzerland	58.6	25	Latvia	50.9
6	Iceland	58.2		South Korea	50.9
7	China	57.8	27	Cyprus	50.8
8	Canada	56.8	28	Austria	50.6
9	Macau	56.6		Czech Republic	50.6
10	Thailand	56.4		Finland	50.6
11	Denmark	54.8	31	Germany	50.5
12	Norway	54.4		United States	50.5
13	Belgium	53.1	33	United Kingdom	50.3
	New Zealand	53.1	34	Azerbaijan	50.0
15	Hong Kong	52.8	35	Ireland	49.8
	Portugal	52.8	36	Slovakia	49.3
17	Brazil	52.4	37	Spain	49.2
18	Netherlands	52.3	38	Ecuador	48.9
19	Japan	52.2	39	Brunei	48.8
20	Sweden	52.1	40	Indonesia	48.5

Most male workforce

Highest % men in workforce

1	Algeria	83.0
2	West Bank and Gaza	82.6
3	Oman	81.6
4	Syria	80.6
5	Pakistan	79.9
6	Iran	79.5
7	Bangladesh	77.7
8	Guatemala	77.4
9	Egypt	77.1
10	Bahrain	76.4
11	Turkey	73.9
12	Tunisia	73.4
13	Panama	73.1
14	Morocco	72.9
15	Fiji	69.3
16	Nicaragua	69.2
17	India	68.4
18	Malta	67.9
19	Sri Lanka	66.7
20	Honduras	65.3
	Malaysia	65.3
22	Chile	64.1
	Mauritius	64.1
24	Paraguay	63.8
25	Indonesia	63.7

Most female workforce

Highest % women in workforce

1	Belarus	53.3
2	Benin	53.1
3	Tanzania	51.0
4	Mongolia	50.7
5	Laos	50.2
6	Estonia	49.9
7	Ghana	49.6
	Madagascar	49.6
9	Armenia	49.5
	Lithuania	49.5
11	Russia	49.4
12	Guadeloupe	49.1
	Moldova	49.1
14	Bahamas	49.0
	Kazakhstan	49.0
16	Barbados	48.7
17	Bermuda	48.5
	Ethiopia	48.5
	Ukraine	48.5
20	Latvia	48.4
21	Zimbabwe	48.2
22	Azerbaijan	48.1
23	Finland	48.0
24	Papua New Guinea	47.9
25	Sweden	47.6

Lowest % of population in labour force

1	West Bank and Gaza	22.8	21	Georgia	38.9
2	Algeria	27.6	22	India	39.1
3	Syria	29.3	23	Mongolia	39.5
4	Egypt	31.0	24	Guinea-Bissau	39.9
5	Pakistan	32.2	25	Panama	40.1
6	Congo-Brazzaville	32.3	26	Malta	40.5
7	Armenia	32.4	27	Sri Lanka	40.6
8	Iran	33.3	28	Fiji	41.1
9	Turkey	34.1	29	El Salvador	41.2
10	Suriname	34.6	30	Malaysia	41.3
	Tunisia	34.6	31	Zimbabwe	41.5
12	Bangladesh	34.7	32	Mexico	41.6
13	Botswana	35.0	33	Croatia	42.2
	Guatemala	35.0		Italy	42.2
15	Morocco	36.4	35	Albania	42.4
16	Nicaragua	36.5	36	Chile	42.6
17	Puerto Rico	37.0	37	Dominican Republic	42.8
18	Oman	37.3	38	Hungary	42.9
19	Moldova	37.6	39	Cuba	43.1
20	Honduras	37.9	40	Macedonia	43.8

Highest rate of unemployment

% of labour force[b]

1	Macedonia	36.0	27	Egypt	11.2
2	Namibia	33.8	28	Croatia	11.1
3	Réunion	29.1		Puerto Rico	11.1
4	Guadeloupe	27.3	30	Uruguay	10.6
5	Guinea-Bissau	26.3	31	Indonesia	10.5
6	South Africa	25.5	32	Germany	10.3
7	Martinique	25.2		Syria	10.3
8	West Bank and Gaza	23.3	34	Turkey	9.9
9	Serbia	20.9	35	Barbados	9.8
10	Dominican Republic	17.9		France	9.8
11	Botswana	17.6	37	Morocco	9.7
12	Ethiopia	16.7	38	Jamaica	9.6
13	Venezuela	15.8	39	Argentina	9.5
14	Netherlands Antilles	15.1	40	Greenland	9.3
15	Tunisia	14.2	41	Mauritius	9.1
16	Burundi	14.0		Panama	9.1
17	Albania	13.8	43	Bulgaria	9.0
	Georgia	13.8	44	Brazil	8.9
	Poland	13.8	45	Greece	8.8
20	Slovakia	13.3		Mali	8.8
21	Jordan	13.2		Peru	8.8
22	Lithuania	12.8	48	Bolivia	8.7
23	Colombia	12.7	49	Afghanistan	8.5
24	Algeria	12.3		Azerbaijan	8.5
25	Nicaragua	12.2		Spain	8.5
26	Iran	11.5			

a Latest available year.   b ILO definition.

The business world

Global competitiveness, 2008

Overall	Government	Infrastructure
1 United States	Singapore	United States
2 Singapore	Hong Kong	Switzerland
3 Hong Kong	Switzerland	Singapore
4 Switzerland	Denmark	Japan
5 Luxembourg	Australia	Sweden
6 Denmark	New Zealand	Germany
7 Australia	Ireland	Denmark
8 Canada	Canada	Canada
9 Sweden	Chile	Netherlands
10 Netherlands	Estonia	Norway
11 Norway	Sweden	France
12 Ireland	China	Finland
13 Taiwan	Finland	Austria
14 Austria	Luxembourg	Israel
15 Finland	Norway	Belgium
16 Germany	Taiwan	Australia
17 China	Netherlands	Taiwan
18 New Zealand	United States	Luxembourg
19 Malaysia	Malaysia	Hong Kong
20 Israel	Austria	United Kingdom
21 United Kingdom	Israel	South Korea
22 Japan	Thailand	New Zealand
23 Estonia	India	Ireland
24 Belgium	United Kingdom	Czech Republic
25 France	Jordan	Malaysia
26 Chile	Germany	Estonia
27 Thailand	Portugal	Hungary
28 Czech Republic	South Africa	Portugal
29 India	Bulgaria	Slovenia
30 Slovakia	Russia	Spain
31 South Korea	Slovakia	China
32 Slovenia	Peru	Lithuania
33 Spain	Czech Republic	Italy
34 Jordan	Spain	Jordan
35 Peru	Colombia	Greece
36 Lithuania	Lithuania	Slovakia
37 Portugal	South Korea	Poland
38 Hungary	Indonesia	Chile
39 Bulgaria	Japan	Thailand
40 Philippines	Mexico	Croatia
41 Colombia	Philippines	Bulgaria
42 Greece	Belgium	Turkey
43 Brazil	Slovenia	Romania
44 Poland	Turkey	Colombia

Notes: Overall competitiveness of 55 economies is calculated by combining four factors: economic performance, government efficiency, business efficiency and infrastructure. Column 1 is based on 331 criteria, using hard data and survey data. Column 2 measures government efficiency, looking at public finance, fiscal policy, institutional and societal frameworks and business legislation. Column 3 includes basic, technological and scientific infrastructure, health and environment, and education.

The business environment

		2008-12 score	2003-2007 score	2003-2007 ranking
1	Denmark	8.76	8.74	2
2	Finland	8.74	8.71	3
3	Singapore	8.73	8.84	1
4	Canada	8.71	8.64	5
5	Switzerland	8.68	8.67	4
6	Australia	8.65	8.20	14
7	Hong Kong	8.64	8.64	7
8	Netherlands	8.63	8.59	8
	Sweden	8.63	8.34	11
10	United States	8.60	8.64	6
11	Ireland	8.57	8.53	9
12	United Kingdom	8.56	8.52	10
13	Germany	8.48	8.13	15
14	New Zealand	8.29	8.20	13
15	Austria	8.26	8.00	16
16	Belgium	8.25	8.22	12
17	Norway	8.14	7.97	17
18	Taiwan	8.11	7.57	22
19	France	8.03	7.89	18
20	Chile	8.02	7.80	79
21	Estonia	7.86	7.79	20
22	Israel	7.85	7.18	25
23	Spain	7.77	7.61	21
24	South Korea	7.59	7.10	29
25	United Arab Emirates	7.57	7.24	24
26	Japan	7.54	7.13	26
27	Qatar	7.53	6.92	31
28	Czech Republic	7.51	7.12	28
29	Slovakia	7.44	6.96	30
30	Portugal	7.41	6.77	33
31	Bahrain	7.40	7.12	27
32	Malaysia	7.39	7.30	23
33	Slovenia	7.37	6.83	32
34	Mexico	7.17	6.71	37
35	Poland	7.12	6.74	35
36	Hungary	7.09	6.73	36
37	Cyprus	7.06	6.74	34
38	Latvia	7.05	6.65	40
39	Italy	7.02	6.58	41
	Lithuania	7.02	6.69	38
41	Brazil	6.98	6.51	42
42	South Africa	6.96	6.13	46
43	Thailand	6.88	6.66	39
44	Greece	6.80	6.44	43
45	Bulgaria	6.77	5.97	47
46	Romania	6.71	5.95	48

Note: Scores reflect the opportunities for, and hindrances to, the conduct of business, measured by countries' rankings in ten categories including market potential, tax and labour-market policies, infrastructure, skills and the political environment. Scores reflect average and forecast average over given date range.

Business creativity and research

Innovation index[a]
2007

1	United States	5.77	13	Netherlands	4.88
2	Switzerland	5.74	14	United Kingdom	4.79
3	Finland	5.67	15	Austria	4.76
4	Japan	5.64	16	Belgium	4.74
5	Israel	5.57	17	France	4.69
6	Sweden	5.53	18	Norway	4.60
7	Germany	5.46	19	Ireland	4.54
8	South Korea	5.36	20	Iceland	4.52
9	Taiwan	5.24	21	Malaysia	4.50
10	Denmark	5.11	22	Australia	4.41
11	Singapore	5.08	23	Hong Kong	4.34
12	Canada	4.90	24	Luxembourg	4.18

Technological readiness index[b]
2007

1	Sweden	5.87	13	Canada	5.34
2	Iceland	5.77	14	Israel	5.29
3	Switzerland	5.67	15	Taiwan	5.27
4	Netherlands	5.65		United Kingdom	5.27
5	Denmark	5.64	17	Australia	5.20
6	Hong Kong	5.48	18	Austria	5.17
7	Norway	5.46	19	Estonia	5.07
	South Korea	5.46	20	Japan	5.06
9	United States	5.43	21	Germany	5.05
10	Luxembourg	5.38	22	France	4.88
11	Finland	5.36	23	Belgium	4.82
	Singapore	5.36		New Zealand	4.82

Brain drain[c]

Highest, 2007			Lowest, 2007		
1	Guyana	1.5	1	United States	6.0
	Lesotho	1.5	2	Qatar	5.7
3	Zimbabwe	1.7	3	United Arab Emirates	5.6
4	Macedonia	2.0	4	Norway	5.5
	Nepal	2.0	5	Chile	5.4
	Serbia	2.0		Ireland	5.4
	Zambia	2.0		Kuwait	5.4
8	Bulgaria	2.1	8	Switzerland	5.3
	Burundi	2.1	9	Finland	5.2
	Moldova	2.1		Japan	5.2
	Senegal	2.1	11	Iceland	5.1

a The innovation index is a measure of the adoption of new technology, and the interaction between the business and science sectors. It includes measures of the investment into research institutions and protection of intellectual property rights.

b The technological readiness index measures the ability of the economy to adopt new technologies. It includes measures of ICT usage, the regulatory framework with regard to ICT, and the availability of new technology to business.

c Scores: 1=talented people leave for other countries, 7=they always remain in home country.

Total expenditure on R&D

% of GDP, 2005			*$bn, 2005*	
1	Israel	4.71	1 United States	312.5
2	Sweden	3.86	2 Japan	146.0
3	Finland	3.48	3 Germany	70.0
4	Japan	3.17	4 United Kingdom	54.8
5	South Korea	2.98	5 France	45.2
6	Switzerland	2.93	6 China	29.9
7	Iceland	2.83	7 South Korea	23.6
8	United States	2.67	8 Canada	22.4
9	United Kingdom	2.55	9 Italy	18.9
10	Taiwan	2.52	10 Sweden	13.8
11	Germany	2.51	11 Spain	12.5
12	Denmark	2.44	12 Australia	11.6
13	Austria	2.36	13 Netherlands	10.8
	Singapore	2.36	14 Switzerland	10.5
15	France	2.13	15 Taiwan	8.7
16	Canada	1.98	16 Russia	8.2
17	Australia	1.82	17 Brazil	7.3
	Belgium	1.82	18 Austria	7.2
19	Netherlands	1.78	19 Finland	6.8
20	Luxembourg	1.56	20 Belgium	6.7
21	Slovenia	1.49	21 Denmark	6.3
22	Norway	1.48	22 Israel	6.1
23	Czech Republic	1.42	23 Norway	4.5
24	China	1.33	24 India	3.7
25	Ireland	1.25	25 Mexico	2.8

Patents

No. of patents granted to residents		*No. of patents in force*	
Total, average 2003–05		*Per 100,000 people, 2005*	
1 Japan	110,714	1 Luxembourg	5,604
2 United States	80,875	2 Taiwan	1,230
3 South Korea	39,650	3 Switzerland	1,152
4 Taiwan	35,599	4 Sweden	1,136
5 Russia	19,699	5 Singapore	991
6 China	16,700	6 Japan	879
7 Germany	13,216	7 South Korea	874
8 Ukraine	9,303	8 Belgium	851
9 France	9,023	9 Ireland	831
10 United Kingdom	3,644	10 New Zealand	830
11 Sweden	2,071	11 United Kingdom	792
12 Spain	1,853	12 Netherlands	776
13 Netherlands	1,827	13 Finland	752
14 Finland	1,170	14 Denmark	701
15 Canada	1,057	15 United States	568
16 Austria	883	16 France	566
17 Poland	815	17 Spain	549
18 Romania	698	18 Germany	525
19 India	695	19 Australia	474
20 Switzerland	626	20 Canada	457

Business costs and FDI

Office occupation costs

Rent, taxes and operating expenses, $ per sq. metre, November 2007

1	London (West End), UK	3,540	13	Dubai, UAE	1,058
2	Mumbai, India	2,040	14	Edinburgh, UK	1,001
3	London (City), UK	1,946	15	Paris (La Défense), France	937
4	Moscow, Russia	1,946	16	Madrid, Spain	905
5	Tokyo (Inner Central), Japan	1,922	17	Oslo, Norway	852
6	Tokyo (Outer Central), Japan	1,663	18	Seoul, South Korea	835
7	Paris, France	1,372	19	Stockholm, Sweden	824
8	New Delhi, India	1,364	20	Zurich, Switzerland	806
9	Dublin, Ireland	1,223	21	Milan, Italy	785
10	Hong Kong	1,144	22	Warsaw, Poland	741
11	Singapore	1,102	23	Abu Dhabi, UAE	733
12	New York (Midtown), US	1,085	24	Frankfurt am Main, Germany	726

Employment costs

Pay, social security and other benefits, production worker, 2007, $ per hour

1	Norway	39.12	11	Canada	26.28
2	Denmark	36.90	12	Australia	26.04
3	Germany	33.49	13	France	25.63
4	Finland	33.01	14	United States	25.26
5	Netherlands	32.27	15	Ireland	23.98
6	Belgium	31.87	16	Italy	21.94
7	Switzerland	30.87	17	Japan	20.72
8	Austria	30.41	18	Spain	18.48
9	Sweden	29.97	19	New Zealand	14.44
10	United Kingdom	27.18	20	Singapore	8.55

Foreign direct investment[a]

	Inflow, 2006, $m			Outflow, 2006, $m	
1	United States	175,394	1	United States	216,614
2	United Kingdom	139,543	2	France	115,036
3	France	81,076	3	Spain	89,679
4	Belgium	71,997	4	Switzerland	81,505
5	China	69,468	5	United Kingdom	79,457
6	Canada	69,041	6	Germany	79,427
7	Hong Kong	42,892	7	Belgium	63,005
8	Germany	42,870	8	Japan	50,266
9	Italy	39,159	9	Canada	45,243
10	Luxembourg	29,309	10	Hong Kong	43,459
11	Russia	28,732	11	Italy	42,035
12	Sweden	27,231	12	Brazil	28,202
13	Switzerland	25,089	13	Sweden	24,600
14	Singapore	24,207	14	Netherlands	22,692
15	Australia	24,022	15	Australia	22,347
16	Turkey	20,120	16	Ireland	22,101
17	Spain	20,016	17	Russia	17,979
18	Mexico	19,037	18	China	16,130

a Investment in companies in a foreign country.

Business burdens and corruption

Number of days taken to register a new company

Highest, 2008		Lowest, 2008	
1 Suriname	694	1 Australia	2
2 Guinea-Bissau	233	2 Canada	3
3 Haiti	202	3 Belgium	4
4 Congo-Kinshasa	155	4 Iceland	5
5 Brazil	152	Singapore	5
6 Venezuela	141	6 Denmark	6
7 Equatorial Guinea	136	Turkey	6
8 Angola	119	United States	6
9 Brunei	116	9 Estonia	7
10 Botswana	108	France	7
11 Indonesia	105	Madagascar	7
12 Laos	103	Mauritius	7
13 Liberia	99	Portugal	7
Namibia	99	Puerto Rico	7
15 Zimbabwe	96	15 Jamaica	8

Corruption perceptions index[a]

2007, 10 = least corrupt

Lowest		Highest	
1 Denmark	9.4	1 Myanmar	1.4
Finland	9.4	Somalia	1.4
New Zealand	9.4	3 Iraq	1.5
4 Singapore	9.3	4 Haiti	1.6
Sweden	9.3	5 Uzbekistan	1.7
6 Iceland	9.2	6 Afghanistan	1.8
7 Netherlands	9.0	Chad	1.8
Switzerland	9.0	Sudan	1.8
9 Canada	8.7	9 Congo-Kinshasa	1.9
Norway	8.7	Equatorial Guinea	1.9
11 Australia	8.6	Guinea	1.9
12 Luxembourg	8.4	Laos	1.9
United Kingdom	8.4	13 Bangladesh	2.0
14 Hong Kong	8.3	Cambodia	2.0
15 Austria	8.1	Central African Rep	2.0
16 Germany	7.8	Papua New Guinea	2.0
17 Ireland	7.5	Turkmenistan	2.0
Japan	7.5	Venezuela	2.0

Business software piracy

% of software that is pirated, 2006

1 Armenia	95	Venezuela	86
2 Azerbaijan	94	8 Indonesia	85
Moldova	94	9 Algeria	84
4 Zimbabwe	91	Cameroon	84
5 Vietnam	88	Ukraine	84
6 Pakistan	86		

a This index ranks countries based on how much corruption is perceived by business
  people, academics and risk analysts to exist among politicians and public officials.

Businesses and banks

Largest businesses
By sales, $bn

1	Wal-Mart Stores	United States	351.1
2	Exxon Mobil	United States	347.3
3	Royal Dutch Shell Group	United Kingdom/Netherlands	318.8
4	BP	United Kingdom	274.3
5	General Motors	United States	207.3
6	Toyota Motor	Japan	204.7
7	ChevronTexaco	United States	200.6
8	DaimlerChrysler	United States	190.2
9	ConocoPhillips	United States	172.5
10	Total Fina Elf	France	168.4
11	General Electric	United States	168.3
12	Ford Motor	United States	160.1
13	ING Group	Netherlands	158.3
14	Citigroup	United States	146.8
15	AXA	France	139.7
16	Volkswagen	Germany	132.3
17	Sinopec	China	131.6
18	Crédit Agricole	France	128.5
19	Allianz	Germany	125.3
20	Fortis	Netherlands	121.2
21	Bank of America Corp	United States	117.0
22	HSBC Holdings	United Kingdom	115.4
23	American Intl. Group	United States	113.2
24	China National Petroleum	China	110.5
25	BNP Paribas	France	109.2
26	ENI	Italy	109.0
27	UBS	Switzerland	107.8
28	Siemens	Germany	107.3
29	State Grid	China	107.2
30	Assicurazioni Generali	Italy	101.8
31	JPMorgan Chase	United States	100.0
32	Carrefour	France	99.0
33	Berkshire Hathaway	United States	98.5
34	Pemex	Mexico	97.7
35	Deutsche Bank	Germany	96.2
36	Dexia Group	Belgium/France	95.8
37	Honda Motor	Japan	94.8
38	McKesson	United States	93.6
39	Verizon Communications	United States	93.2
40	Nippon Telegraph & Telephone	Japan	92.0
41	Hewlett-Packard	United States	91.7
42	IBM	United States	91.4
43	Valero Energy	United States	91.1
44	Home Depot	United States	90.8

Notes: Industrial and service corporations. Figures refer to the year ended December 31, 2006, except for Japanese companies, where figures refer to year ended March 31, 2007. They include sales of consolidated subsidiaries but exclude excise taxes, thus differing, in some instances, from figures published by the companies themselves.

Largest banks

By capital, $m

1	Bank of America Corp	United States	91,065
2	Citigroup	United States	90,899
3	HSBC Holdings	United Kingdom	87,842
4	Crédit Agricole Groupe	France	84,937
5	JPMorgan Chase	United States	81,055
6	Mitsubishi UFJ Financial Group	Japan	68,464
7	Industrial and Commercial Bank of China	China	59,166
8	Royal Bank of Scotland	United Kingdom	58,973
9	Bank of China	China	52,518
10	Santander Central Hispano	Spain	46,805
11	BNP Paribas	France	45,605
12	Barclays Bank	United Kingdom	45,161
13	HBOS	United Kingdom	44,030
14	China Construction Bank	China	42,286
15	Mizuho Financial Group	Japan	41,934
16	Wachovia Corporation	United States	39,428
17	UniCredit	Italy	38,700
18	Wells Fargo & Co.	United States	36,808
19	Rabobank Group	Netherlands	34,757
20	ING Bank	Netherlands	33,358
21	UBS	Switzerland	33,212
22	Sumitomo Mitsui Financial Group	Japan	33,177
23	Deutsche Bank	Germany	32,264
24	ABN-Amro Bank	Netherlands	31,239
25	Crédit Mutuel	France	29,792
26	Société Générale	France	29,405
27	Credit Suisse Group	Switzerland	28,805
28	Banco Bilbao Vizcaya Argentaria	Spain	25,779
29	Lloyds TSB Group	United Kingdom	25,185
30	Groupe Caisse d'Epargne	France	24,159
31	Groupe Banques Populaires	France	22,257
32	Fortis Bank	Belgium	22,255
33	Norinchukin Bank	Japan	21,194
34	Commerzbank	Germany	20,410
35	Royal Bank of Canada	Canada	19,131
36	Washington Mutual	United States	17,919
37	Scotiabank	Canada	17,911
38	National Australia Bank	Australia	17,506
39	Nordea Group	Sweden	17,315
40	Dexia	Belgium	17,158
41	US Bancorp	United States	17,036
42	Danske Bank	Denmark	16,988
43	Banca Intesa	Italy	16,736
44	Dresdner Bank	Germany	16,422
45	Caja de Ahorros y Pen.de Barcelona	Spain	15,797
46	Sanpaolo IMI	Italy	15,589

Notes: Capital (tier one) is essentially equity and reserves.
Figures for Japanese banks refer to the year ended March 31, 2007. Figures for all other countries refer to the year ended December 31, 2006.

Stockmarkets

Largest market capitalisation
$bn, end 2007

1	United States	19,947	27	Malaysia	326
2	China	6,226	28	Turkey	287
3	Japan	4,453	29	Denmark	278
4	United Kingdom	3,859	30	Greece	265
5	France	2,771	31	Israel	236
6	Canada	2,187	32	Austria	229
7	Germany	2,106	33	United Arab Emirates	225
8	India	1,819	34	Chile	213
9	Spain	1,800	35	Indonesia	212
10	Russia	1,503	36	Poland	207
11	Brazil	1,370	37	Thailand	196
12	Australia	1,298	38	Kuwait	188
13	Switzerland	1,275	39	Luxembourg	166
14	Hong Kong	1,163	40	Ireland	144
15	South Korea	1,124	41	Egypt	139
16	Italy	1,073	42	Portugal	132
17	Netherlands	956	43	Ukraine	112
18	South Africa	834	44	Peru	106
19	Taiwan	724	45	Philippines	103
20	Sweden	612	46	Colombia	102
21	Saudi Arabia	515	47	Qatar	95
22	Mexico	398	48	Argentina	87
23	Belgium	386	49	Nigeria	86
24	Finland	369	50	Morocco	75
25	Norway	357	51	Czech Republic	73
26	Singapore	353	52	Pakistan	70

Highest growth in market capitalisation, $ terms
% increase, 2002–07

1	Vietnam[a]	12,590	21	Montenegro[c]	942
2	Ukraine	3,483	22	Zambia	907
3	Serbia	3,161	23	Romania	885
4	Kazakhstan	2,986	24	Kenya	841
5	Bulgaria	2,873	25	Morocco	779
6	Uzbekistan[b]	2,206	26	Turkey	744
7	Mongolia	1,813	27	Peru	693
8	Kyrgyzstan	1,629	28	Lebanon	675
9	Croatia	1,559	29	Poland	621
10	Nigeria	1,404	30	Austria	617
11	Macedonia	1,392	31	Indonesia	606
12	India	1,289	32	Lithuania	593
13	China	1,245	33	Pakistan	589
14	Georgia	1,223	34	Saudi Arabia	588
15	Russia	1,110	35	Malawi[a]	583
16	Nepal	1,077	36	Luxembourg	572
17	Brazil	1,007	37	Iceland	541
18	United Arab Emirates	1,003	38	Côte d'Ivoire	529
19	Armenia	960		Slovenia	529
20	Colombia	955	40	Kuwait	512

a 2003–07   b 2002–06   c 2004–07

Highest growth in value traded

$ terms, % increase, 2002–07

1	Vietnam[a]	78,544	23	Ecuador	1,437
2	United Arab Emirates	41,316	24	Poland	1,348
3	Kazakhstan	6,349	25	Iceland	1,344
4	Mongolia	5,200	26	Chile	1,325
5	Morocco	4,376	27	Jordan	1,203
6	Colombia	3,687	28	Papua New Guinea[b]	1,150
7	Kenya	3,561	29	Montenegro[a]	1,150
8	Zambia	3,500	30	Brazil	1,114
9	Nigeria	3,431	31	Nepal	1,055
10	Bulgaria	3,097	32	Georgia	1,025
11	Croatia	2,681	33	Oman	922
12	China	2,237	34	Ghana	891
13	Namibia	2,200	35	Côte d'Ivoire	881
14	Serbia	2,198	36	Norway	865
15	Austria	1,988	37	Philippines	843
	Russia	1,988	38	Estonia	770
17	Egypt	1,975	39	Indonesia	765
18	Romania	1,909	40	Cyprus	758
19	Saudi Arabia	1,806	41	Lebanon	733
20	West Bank and Gaza	1,707	42	Hungary	699
21	Macedonia	1,638	43	Bahrain	667
22	Ukraine	1,488	44	Belgium	656

Highest growth in number of listed companies

% increase, 2002–07

1	Uzbekistan	1,800.0	25	Ghana	33.3
2	Serbia	1,688.9	26	Oman	30.2
3	Vietnam[a]	450.0	27	West Bank and Gaza	29.6
4	Croatia	434.8	28	Argentina	28.9
5	United Arab Emirates	275.0	29	Estonia	28.6
6	Slovenia	148.6	30	Japan	25.7
7	Mauritius	125.0	31	Malta	25.0
8	Kuwait	112.9	32	China	23.9
9	Taiwan	95.6	33	Bolivia	23.3
10	Papua New Guinea	87.5	34	Swaziland	20.0
11	Kazakhstan	83.7	35	Malaysia	19.8
12	Russia	67.3	36	Trinidad & Tobago	19.4
13	El Salvador	64.5	37	Thailand	19.3
14	Saudi Arabia	63.2	38	New Zealand	17.6
15	Montenegro[c]	60.0	39	Spain	17.1
16	Jordan	55.1	40	South Korea	16.4
17	Poland	51.9	41	Bangladesh	16.3
18	Ukraine	50.0	42	Norway	16.2
19	Nepal	46.9	43	Indonesia	15.7
20	Australia	41.2	44	Belgium	14.0
21	Tanzania	40.0	45	Ecuador	12.9
22	Qatar[d]	37.9	46	Malawi[a]	12.5
23	Zambia	36.4	47	Austria	12.1
24	Morocco	34.5			

a 2003–07 b 2002–06 c 2003–06 d 2004–07

Transport: roads and cars

Longest road networks
Km, 2005 or latest

1	United States	6,544,257	21	Germany	231,480
2	India	3,383,344	22	Argentina	231,374
3	China	1,930,544	23	Vietnam	222,179
4	Brazil	1,751,868	24	Philippines	200,037
5	Canada	1,408,900	25	Romania	198,817
6	Japan	1,177,278	26	Nigeria	193,200
7	France	950,985	27	Iran	179,388
8	Australia	812,972	28	Ukraine	169,323
9	Spain	666,292	29	Colombia	164,257
10	Russia	537,289	30	Hungary	159,568
11	Italy	484,688	31	Congo-Kinshasa	153,497
12	Turkey	426,914	32	Saudi Arabia	152,044
13	Sweden	425,383	33	Belgium	150,567
14	Poland	423,997	34	Austria	133,901
15	United Kingdom	388,008	35	Czech Republic	127,781
16	Indonesia	368,360	36	Netherlands	126,100
17	South Africa	364,131	37	Greece	117,533
18	Mexico	355,796	38	Algeria	108,302
19	Pakistan	258,340	39	South Korea	102,293
20	Bangladesh	239,226	40	Malaysia	98,721

Densest road networks
Km of road per km^2 land area, 2005 or latest

1	Macau	21.3		Trinidad & Tobago	1.6
2	Malta	7.1		United Kingdom	1.6
3	Bahrain	5.1	24	Sri Lanka	1.5
	Singapore	5.1	25	Ireland	1.4
5	Belgium	4.9		Poland	1.4
6	Barbados	3.7	27	Cyprus	1.3
7	Japan	3.1		Estonia	1.3
8	Netherlands	3.0		Spain	1.3
9	Puerto Rico	2.8	30	Lithuania	1.2
10	Luxembourg	2.0	31	Latvia	1.1
	Jamaica	2.0		Netherlands Antilles	1.1
12	Slovenia	1.9	33	India	1.0
13	Hong Kong	1.8		Mauritius	1.0
14	Bangladesh	1.7		South Korea	1.0
	Denmark	1.7	36	Greece	0.9
	France	1.7		Portugal	0.9
	Hungary	1.7		Slovakia	0.9
	Switzerland	1.7		Sweden	0.9
19	Austria	1.6	40	Israel	0.8
	Czech Republic	1.6		Romania	0.8
	Italy	1.6			

Most crowded road networks

Number of vehicles per km of road network, 2005 or latest

1	Qatar	283.6	27	Switzerland	58.8
2	Hong Kong	252.5	28	Croatia	54.5
3	Germany	208.3	29	Guatemala	52.7
4	Macau	185.5	30	Tunisia	49.1
5	Singapore	183.4	31	Russia	47.3
6	Kuwait	180.7	32	Greece	46.9
7	South Korea	141.8	33	Cyprus	46.6
8	Israel	115.2	34	France	38.2
9	Malta	113.2	35	Belgium	36.6
10	Serbia	102.1	36	Austria	36.5
11	Netherlands Antilles	95.3	37	Ukraine	35.6
12	Jordan	82.9	38	Brunei	35.5
13	Dominican Republic	81.4	39	Finland	35.3
14	Italy	80.7	40	Poland	34.7
15	Mauritius	79.4		Spain	34.7
16	United Kingdom	78.7	42	Denmark	32.7
17	Bahrain	77.2	43	Slovakia	32.1
18	Malaysia	67.4	44	New Zealand	31.6
	Portugal	67.4	45	Czech Republic	31.4
20	Japan	63.6	46	Moldova	31.0
21	Bulgaria	63.4	47	Honduras	30.7
22	Luxembourg	63.3	48	United States	30.6
23	Barbados	63.1	49	Morocco	29.5
24	Indonesia	62.4	50	Algeria	27.5
25	Netherlands	62.2		Norway	27.5
26	Mexico	60.3			

Most used road networks

'000 vehicle-km per year per km of road network, 2005 or latest

1	Hong Kong	5,531.5	21	France	575.7
2	Singapore	4,261.0	22	Ecuador	568.1
3	Germany	2,760.5	23	Bulgaria	521.7
4	Israel	2,335.5	24	China	435.6
5	Bahrain	1,528.0	25	New Zealand	434.5
6	South Korea	1,288.6	26	Norway	381.3
7	United Kingdom	1,271.7	27	Czech Republic	369.1
8	Tunisia	999.9	28	Austria	357.1
9	Netherlands	872.0	29	Ireland	351.1
10	Switzerland	839.7	30	Romania	341.2
11	Cyprus	783.5	31	South Africa	338.9
12	Luxembourg	778.5	32	Spain	336.7
13	United States	732.6	33	Poland	325.7
14	Greece	675.0	34	Guatemala	322.6
15	Japan	664.0	35	Ghana	320.6
16	Finland	655.2		Suriname	320.6
17	Croatia	635.9	37	Egypt	310.6
18	Peru	633.6	38	Morocco	309.5
19	Denmark	629.8	39	Senegal	295.6
20	Belgium	621.0	40	Slovenia	288.6

Highest car ownership
Number of cars per 1,000 population, 2005 or latest

1	Luxembourg	647	27	Greece	388
2	Iceland	632	28	Ireland	382
3	New Zealand	607	29	Estonia	367
4	Italy	595	30	Netherlands Antilles	360
5	Canada	561	31	Czech Republic	358
6	Cyprus	550	32	Denmark	354
	Germany	550	33	Kuwait	349
8	Australia	542	34	Brunei	346
9	Malta	523	35	Barbados	343
10	Switzerland	520	36	Qatar	335
11	Austria	503	37	Bahrain	325
12	France	494	38	Latvia	323
13	Portugal	471		Poland	323
	Slovenia	471	40	Bulgaria	314
15	Belgium	468	41	Croatia	312
16	United States	461	42	Bahamas	290
17	Finland	460	43	Hungary	274
	Sweden	460	44	Israel	239
19	United Kingdom	451	45	Libya	232
20	Spain	445	46	South Korea	230
21	Japan	441	47	United Arab Emirates	228
22	Norway	439	48	Malaysia	225
23	Lithuania	426	49	Slovakia	222
24	Netherlands	424	50	Belarus	181
25	Saudi Arabia	415		Serbia	181
26	Lebanon	403			

Lowest car ownership
Number of cars per 1,000 population, 2005 or latest

1	Bangladesh	1		Guinea	8
	Burundi	1		India	8
	Central African Rep	1	24	Kenya	9
	Ethiopia	1		Philippines	9
	Rwanda	1	26	Pakistan	10
	Tanzania	1		Senegal	10
7	Sierra Leone	2		Togo	10
	Uganda	2	29	Bhutan	12
9	Equatorial Guinea	3		Syria	12
	Mali	3	31	Benin	13
	Nepal	3		Sri Lanka	13
12	Myanmar	4	33	Bolivia	15
	Niger	4		China	15
14	Burkina Faso	5	35	Nigeria	17
	Gambia, The	5	36	Nicaragua	18
	Papua New Guinea	5	37	Tajikistan	19
17	Liberia	6		Yemen	19
18	Côte d'Ivoire	7	39	El Salvador	24
19	Angola	8		Iran	24
	Cameroon	8	41	Cambodia	25
	Congo-Kinshasa	8	42	Egypt	27

Most injured in road accidents
Number of people injured per 100,000 population, 2005 or latest

1	Qatar	9,989	26	Germany	408
2	Kuwait	2,231	27	Oman	406
3	Mauritius	1,580	28	Italy	390
4	Jordan	1,519	29	Portugal	387
5	Saudi Arabia	1,305	30	Croatia	353
6	Costa Rica	1,231	31	Cyprus	336
7	Malaysia	1,222	32	United Kingdom	330
8	Panama	1,212	33	Malta	298
9	Botswana	1,025	34	Switzerland	292
10	Suriname	913	35	Sri Lanka	290
11	Bosnia	867	36	Chile	287
12	Serbia	771	37	South Africa	281
13	Barbados	769	38	Mongolia	275
14	Japan	745	39	Peru	271
15	Turkey	741	40	New Zealand	264
16	Brunei	734	41	Czech Republic	260
17	United States	626	42	Israel	245
18	Austria	522		Nicaragua	245
19	Slovenia	515	44	Bahrain	237
20	Colombia	510	45	Bolivia	230
21	Swaziland	501	46	Iceland	226
22	Canada	473	47	Spain	220
23	Belgium	467	48	Hong Kong	217
24	South Korea	443	49	Lesotho	215
25	Namibia	440	50	Hungary	207

Most deaths in road accidents
Number of people killed per 100,000 population, 2005 or latest

1	South Africa	31	21	Greece	15
2	Botswana	30		Suriname	15
3	Malaysia	26		Tunisia	15
4	Russia	24		Ukraine	15
	Swaziland	24		United States	15
6	Gabon	23	26	Czech Republic	14
	Lithuania	23		Jordan	14
	Oman	23		Poland	14
	Thailand	23		Vietnam	14
10	Kazakhstan	22	30	Algeria	13
	United Arab Emirates	22		Azerbaijan	13
12	Saudi Arabia	21		Croatia	13
13	Latvia	19		Cyprus	13
14	Belarus	17		Georgia	13
	Kyrgyzstan	17		Hungary	13
	Mongolia	17		Panama	13
	Namibia	17		Puerto Rico	13
	Qatar	17		Slovenia	13
19	Kuwait	16		South Korea	13
	Lesotho	16			

Transport: planes and trains

Most air travel

Million passenger-km[a] per year, 2007

1	United States	1,271,344	16	Russia	51,884
2	China	228,484	17	Italy	47,465
3	United Kingdom	218,967	18	Malaysia	38,956
4	Germany	144,005	19	Mexico	32,813
5	France	123,336	20	Saudi Arabia	29,715
6	Singapore	90,126	21	Indonesia	28,949
7	Australia	77,739	22	South Africa	28,760
8	Spain	77,265	23	New Zealand	27,592
9	South Korea	72,823	24	Ireland	25,484
10	Canada	72,486	25	Belgium	25,028
11	Netherlands	71,771	26	Turkey	23,301
12	Hong Kong	70,592	27	Austria	20,072
13	United Arab Emirates	65,491	28	Switzerland	19,434
14	India	59,269	29	Portugal	19,002
15	Thailand	55,292	30	Israel	18,067

Busiest airports

Total passengers, m, 2007

1	Atlanta, Hartsfield	89.4
2	Chicago, O'Hare	76.2
3	London, Heathrow	68.1
4	Tokyo, Haneda	66.7
5	Los Angeles, Intl.	61.9
6	Dallas, Ft. Worth	59.8
7	Paris, Charles de Gaulle	59.9
8	Frankfurt, Main	54.2
9	Beijing, Capital	53.7
10	Madrid, Barajas	52.1
11	Denver, Intl.	49.9
12	New York, JFK	47.8

Total cargo, m tonnes, 2007

1	Memphis, Intl.	3.84
2	Hong Kong, Intl.	3.77
3	Anchorage, Intl.	2.83
4	Seoul, Inchon	2.56
5	Shanghai, Pudong Intl.	2.49
6	Paris, Charles de Gaulle	2.30
7	Tokyo, Narita	2.25
8	Frankfurt, Main	2.17
9	Louisville, Standiford Fd.	2.08
10	Miami, Intl.	1.92
	Singapore, Changi	1.92
12	Los Angeles, Intl.	1.88

Average daily aircraft movements, take-offs and landings, 2007

1	Atlanta, Hartsfield	2,724
2	Chicago, O'Hare	2,542
3	Dallas, Ft. Worth	1,876
4	Los Angeles, Intl.	1,867
5	Denver, Intl.	1,683
6	Las Vegas, McCarran Intl.	1,670
7	Houston, George Bush Intercont.	1,654
8	Paris, Charles de Gaulle	1,514
9	Phoenix, Skyharbor Intl.	1,474
10	Charlotte/ Douglas, Intl.	1,432
11	Philadelphia, Intl.	1,367
12	Frankfurt, Main	1,350
13	Madrid, Barajas	1,324
14	London, Heathrow	1,319
15	Detroit, Metro	1,280
16	Amsterdam, Schipol	1,245
17	Minneapolis, St Paul	1,234
18	Newark	1,216
19	New York, JFK	1,214
20	Munich, Intl.	1,183

a Air passenger–km data refer to the distance travelled by aircraft of national origin.

Longest railway networks
'000 km, 2006

1	United States	227.2		21	Australia	9.6
2	Russia	92.2		22	Turkey	8.7
3	China	63.4		23	Hungary	7.9
4	India	63.3		24	Pakistan	7.8
5	Canada	57.5		25	Iran	7.3
6	Argentina	35.8		26	Chile	5.9
7	Germany	34.1			Finland	5.9
8	France	29.5		28	Austria	5.8
9	Brazil	29.3		29	Belarus	5.5
10	Mexico	26.7			Sudan	5.5
11	Ukraine	22.0		31	Egypt	5.2
12	Japan	20.1		32	Bulgaria	4.1
13	South Africa	20.0		33	Norway	4.0
14	United Kingdom	19.9			Uzbekistan	4.0
15	Poland	19.4		35	Serbia	3.8
16	Italy	16.8		36	Slovakia	3.7
17	Spain	14.6		37	Algeria	3.6
18	Kazakhstan	14.2			Belgium	3.6
19	Romania	10.8			Congo-Brazzaville	3.6
20	Sweden	10.0			Nigeria	3.6

Most rail passengers
Km per person per year, 2006

1	Switzerland	2,123		11	Kazakhstan	908
2	Japan	1,946		12	Belgium	873
3	France	1,286		13	Italy	807
4	Russia	1,223		14	United Kingdom	766
5	Ukraine	1,133		15	Finland	721
6	Denmark	1,130		16	Hungary	692
7	Austria	1,096		17	Czech Republic	689
8	Belarus	997		18	South Korea	655
9	Netherlands	963		19	Sweden	631
10	Germany	911		20	Egypt	559

Most rail freight
Million tonnes-km per year, 2006

1	United States	2,839,124		11	Mexico	54,387
2	China	2,055,716		12	Australia	46,036
3	Russia	1,950,000		13	Belarus	45,723
4	India	439,596		14	Poland	42,651
5	Canada	352,069		15	France	42,124
6	Ukraine	240,810		16	Japan	23,014
7	Brazil	232,300		17	United Kingdom	22,180
8	Kazakhstan	191,189		18	Italy	21,852
9	South Africa	108,513		19	Iran	20,542
10	Germany	86,690		20	Austria	19,525

Transport: shipping

Merchant fleets

Number of vessels, by country of domicile, 2006

1	Japan	3,330	11	Singapore	794
2	China	3,184	12	Indonesia	793
3	Greece	3,084	13	Denmark	781
4	Germany	2,964	14	Italy	739
5	Russia	2,203		Netherlands	739
6	Norway	1,810	16	Hong Kong	689
7	United States	1,766	17	Taiwan	574
8	South Korea	1,041	18	India	456
9	Turkey	874	19	Switzerland	370
10	United Kingdom	855	20	United Arab Emirates	366

By country of domicile, deadweight tonnage, m, 2006

1	Greece	170.2	11	Taiwan	24.9
2	Japan	147.5	12	Denmark	21.9
3	Germany	85.0	13	Russia	18.1
4	China	70.4	14	Italy	16.0
5	Norway	48.7	15	India	14.8
6	United States	48.3	16	Switzerland	12.5
7	Hong Kong	45.1	17	Belgium	12.5
8	South Korea	32.3	18	Saudi Arabia	11.9
9	United Kingdom	26.8	19	Turkey	10.9
10	Singapore	25.7	20	Iran	10.0

Maritime trading

% of value of world trade generated, 2006

1	United States	12.2	14	Mexico	2.1
2	Germany	8.5		Singapore	2.1
3	China	7.3	16	Russia	1.8
4	Japan	5.1		Taiwan	1.8
5	United Kingdom	4.3	18	Austria	1.2
6	France	4.2		India	1.2
7	Netherlands	3.6		Malaysia	1.2
8	Italy	3.5	21	Australia	1.1
9	Canada	3.2		Saudi Arabia	1.1
10	Belgium	3.0		Sweden	1.1
11	Hong Kong	2.7		Switzerland	1.1
12	South Korea	2.6		United Arab Emirates	1.1
13	Spain	2.2			

% of world fleet deadweight tonnage, by country of ownership, 2006

1	Japan	15.1	11	Italy	1.6
2	Germany	8.7	12	India	1.5
3	China	7.2	13	Austria	1.3
4	United States	4.9		Belgium	1.3
5	Hong Kong	4.6		Switzerland	1.3
6	South Korea	3.3	16	Saudi Arabia	1.2
7	United Kingdom	2.7	17	Netherlands	0.9
8	Singapore	2.6	18	Malaysia	0.7
9	Taiwan	2.5		Sweden	0.7
10	Russia	1.9		United Arab Emirates	0.7

Tourism

Most tourist arrivals
Number of arrivals, '000, 2006

1	France	79,083	21	Macau	10,683
2	Spain	58,451	22	Croatia	8,659
3	United States	51,063	23	Egypt	8,646
4	China	49,600	24	Saudi Arabia	8,620
5	Italy	41,058	25	South Africa	8,396
6	United Kingdom	30,654	26	Hungary	8,259
7	Germany	23,569	27	Ireland	8,001
8	Mexico	21,353	28	Switzerland	7,863
9	Austria	20,261	29	Singapore	7,588
10	Russia	20,199	30	Japan	7,334
11	Ukraine	18,936	31	Belgium	6,995
12	Turkey	18,916	32	Morocco	6,558
13	Canada	18,265	33	Czech Republic	6,435
14	Malaysia	17,547	34	Tunisia	6,378
15	Greece	16,039	35	South Korea	6,155
16	Hong Kong	15,821	36	Bulgaria	5,158
17	Poland	15,670	37	Australia	5,064
18	Thailand	13,882	38	Brazil	5,019
19	Portugal	11,282	39	Indonesia	4,871
20	Netherlands	10,739	40	Bahrain	4,519

Biggest tourist spenders
$m, 2006

1	United States	74,807	11	South Korea	15,732
2	Germany	73,765	12	Belgium	15,619
3	United Kingdom	61,929	13	Spain	15,463
4	Japan	38,031	14	Hong Kong	13,359
5	France	32,868	15	Australia	11,995
6	Italy	22,991	16	Austria	11,334
7	China	22,950	17	Sweden	11,069
8	Canada	19,315	18	Singapore	10,020
9	Russia	18,081	19	Norway	9,954
10	Netherlands	16,377	20	Switzerland	9,460

Largest tourist receipts
$m, 2006

1	United States	85,694	11	Canada	14,632
2	Spain	51,115	12	Greece	14,259
3	France	42,910	13	Thailand	12,423
4	Italy	38,129	14	Mexico	12,177
5	China	33,949	15	Switzerland	11,843
6	United Kingdom	33,695	16	Hong Kong	11,630
7	Germany	32,760	17	Belgium	11,535
8	Australia	17,840	18	Netherlands	11,516
9	Turkey	16,853	19	Malaysia	9,630
10	Austria	16,658	20	Macau	9,337

Education

Primary enrolment
Number enrolled as % of relevant age group

Highest			Lowest		
1	Gabon	152	1	Niger	51
2	Sierra Leone	147	2	Papua New Guinea	55
3	Rwanda	140	3	Burkina Faso	60
4	Madagascar	139	4	Central African Rep	61
5	Brazil	137		Congo-Brazzaville	61
6	Nepal	126	6	Eritrea	62
	Syria	126	7	Sudan	66
8	Netherlands Antilles	124	8	Côte d'Ivoire	71
9	Belize	123	9	Chad	76
10	Cambodia	122	10	Mali	80
	Equatorial Guinea	122		Senegal	80
12	Suriname	121	12	Oman	82
13	Bahrain	120	13	Pakistan	84
14	Malawi	119	14	Yemen	87

Highest tertiary enrolment[a]
Number enrolled as % of relevant age group

1	Greece	95		Iceland	73
2	Finland	93		Ukraine	73
	South Korea	93	16	Russia	72
4	Cuba	88	17	Hungary	69
5	Slovenia	83	18	Italy	67
6	United States	82		Spain	67
7	Denmark	80	20	Belarus	66
	New Zealand	80		Poland	66
9	Sweden	79	22	Estonia	65
10	Norway	78	23	Argentina	64
11	Lithuania	76	24	Belgium	63
12	Latvia	74	25	Netherlands	60
13	Australia	73			

Education spending
% of GDP

Highest			Lowest		
1	Lesotho	13.0	1	Equatorial Guinea	0.6
2	Cuba	9.1	2	Bermuda	1.2
3	Botswana	8.7	3	Guinea	1.6
4	Denmark	8.3	4	Cambodia	1.7
5	Iceland	7.6	5	Chad	1.9
	Moldova	7.6		Congo-Kinshasa	1.9
7	Tunisia	7.3	7	Gambia, The	2.0
8	Norway	7.2		Zambia	2.0

Notes: Latest available year 2003–07. The gross enrolment ratios shown are the actual number enrolled as a percentage of the number of children in the official primary age group. They may exceed 100 when children outside the primary age group are receiving primary education.

a Tertiary education includes all levels of post-secondary education including courses leading to awards not equivalent to a university degree, courses leading to a first university degree and postgraduate courses.

Least literate
% adult literacy rate[a]

1	Mali	23.3	25	Burundi	59.3
2	Chad	25.7	26	Sudan	60.9
3	Afghanistan	28.0	27	Haiti	62.1
4	Burkina Faso	28.7	28	Guinea-Bissau	64.6
5	Guinea	29.5	29	Rwanda	64.9
6	Niger	30.4	30	Ghana	65.0
7	Ethiopia	35.9	31	India	66.0
8	Sierra Leone	38.1	32	Congo-Kinshasa	67.2
9	Benin	40.5	33	Angola	67.4
10	Senegal	42.6	34	Cameroon	67.9
11	Mozambique	44.4	35	Madagascar	70.7
12	Central African Rep	48.6	36	Malawi	71.8
13	Côte d'Ivoire	48.7	37	Egypt	72.0
14	Togo	53.2		Nigeria	72.0
15	Bangladesh	53.5	39	Tanzania	72.3
16	Morocco	54.7	40	Guatemala	73.2
17	Pakistan	54.9		Laos	73.2
18	Liberia	55.5	42	Kenya	73.6
19	Bhutan	55.6		Uganda	73.6
20	Mauritania	55.8	44	Iraq	74.1
21	Nepal	56.5	45	Algeria	75.4
22	Yemen	57.3	46	Cambodia	76.3
23	Papua New Guinea	57.8	47	Tunisia	77.7
24	Yemen	58.9	48	Swaziland	79.6

Student performance

Reading score, PISA 2006 results[b]

1	South Korea	556
2	Finland	547
3	Hong Kong	536
4	Canada	527
5	New Zealand	521
6	Ireland	517
7	Australia	513
8	Poland	508
9	Netherlands	507
	Sweden	507
11	Belgium	501
	Estonia	501

Maths score, PISA 2006 results[b]

1	Finland	548
2	Hong Kong	547
	South Korea	547
4	Netherlands	531
5	Switzerland	530
6	Canada	527
7	Macau	525
8	Japan	523
9	New Zealand	522
10	Australia	520
	Belgium	520
12	Estonia	515

Science score, PISA 2006 results[b]

1	Finland	563
2	Hong Kong	542
3	Canada	534
4	Estonia	531
	Japan	531
6	New Zealand	530

7	Australia	527
8	Netherlands	525
9	South Korea	522
10	Slovenia	519
11	Germany	516
12	United Kingdom	515

a Latest available year 2000–07.
b PISA is the OECD's Programme for International Student Assessment, a survey every three years of 15 year-olds in the principal industrialised countries. The average score for the countries covered is 500.

Life expectancy

Highest life expectancy
Years, 2005–10

1	Andorra[a]	83.5		Malta	79.4
2	Japan	82.6		United Kingdom	79.4
3	Hong Kong	82.2		Virgin Islands (US)	79.4
4	Iceland	81.8	28	Finland	79.3
5	Switzerland	81.7	29	Guadeloupe	79.2
6	Australia	81.2	30	Channel Islands	79.0
7	Spain	80.9		Cyprus	79.0
	Sweden	80.9	32	Ireland	78.9
9	Canada	80.7	33	Costa Rica	78.8
	France	80.7	34	Luxembourg	78.7
	Israel	80.7		Puerto Rico	78.7
	Macau	80.7		United Arab Emirates	78.7
13	Italy	80.5	37	Chile	78.6
14	Cayman Islands[a]	80.2		South Korea	78.6
	New Zealand	80.2	39	Cuba	78.3
	Norway	80.2		Denmark	78.3
17	Singapore	80.0	41	United States	78.2
18	Austria	79.8	42	Bermuda[a]	78.1
	Netherlands	79.8		Portugal	78.1
20	Faroe Islands[a]	79.5	44	Slovenia	77.9
	Greece	79.5	45	Kuwait	77.6
	Martinique	79.5		Taiwan[a]	77.6
23	Belgium	79.4	47	Barbados	77.3
	Germany	79.4	48	Brunei	77.1

Highest male life expectancy
Years, 2005–10

1	Andorra[a]	80.6	10	Canada	78.3
2	Iceland	80.2	11	New Zealand	78.2
3	Hong Kong	79.4	12	Singapore	78.0
4	Japan	79.0	13	Norway	77.8
	Switzerland	79.0	14	Spain	77.7
6	Australia	78.9	15	Cayman Islands[a]	77.6
7	Sweden	78.7	16	Italy	77.5
8	Israel	78.6		Netherlands	77.5
9	Macau	78.5	18	Malta	77.3

Highest female life expectancy
Years, 2005–10

1	Andorra[a]	86.6		Virgin Islands (US)	83.3
2	Japan	86.1	11	Sweden	83.0
3	Hong Kong	85.1	12	Canada	82.9
4	Spain	84.2		Cayman Islands[a]	82.9
	Switzerland	84.2		Faroe Islands[a]	82.9
6	France	84.1	15	Israel	82.8
7	Australia	83.6		Macau	82.8
8	Italy	83.5	17	Puerto Rico	82.7
9	Iceland	83.3	18	Austria	82.6

a 2007 estimate.

Lowest life expectancy
Years, 2005–10

1	Swaziland	39.6	26	Tanzania	52.5
2	Mozambique	42.1	27	Ethiopia	52.9
3	Zambia	42.4		Namibia	52.9
4	Lesotho	42.6	29	Kenya	54.1
	Sierra Leone	42.6	30	Mali	54.5
6	Angola	42.7	31	Congo-Brazzaville	55.3
7	Zimbabwe	43.5	32	Guinea	56.0
8	Afghanistan	43.8	33	Benin	56.7
9	Central African Rep	44.7		Gabon	56.7
10	Liberia	45.7	35	Niger	56.9
11	Rwanda	46.2	36	Papua New Guinea	57.2
12	Guinea-Bissau	46.4	37	Eritrea	58.0
13	Congo-Kinshasa	46.5	38	Togo	58.4
14	Nigeria	46.9	39	Sudan	58.6
15	Somalia	48.2	40	Gambia, The	59.4
16	Côte d'Ivoire	48.3		Madagascar	59.4
	Malawi	48.3	42	Iraq	59.5
18	South Africa	49.3	43	Cambodia	59.7
19	Burundi	49.6	44	Ghana	60.0
20	Cameroon	50.4	45	Timor-Leste	60.8
21	Botswana	50.7	46	Haiti	60.9
	Chad	50.7	47	Myanmar	62.1
23	Uganda	51.5	48	Yemen	62.7
24	Equatorial Guinea	51.6	49	Senegal	63.1
25	Burkina Faso	52.3	50	Turkmenistan	63.2

Lowest male life expectancy
Years, 2005–10

1	Swaziland	39.8	11	Liberia	44.8
2	Sierra Leone	41.0	12	Guinea-Bissau	44.9
3	Angola	41.2	13	Congo-Kinshasa	45.2
4	Mozambique	41.7	14	Nigeria	46.4
5	Zambia	42.1	15	Somalia	46.9
6	Lesotho	42.9	16	Côte d'Ivoire	47.5
7	Central African Rep	43.3	17	Burundi	48.1
8	Afghanistan	43.9		Malawi	48.1
9	Zimbabwe	44.1	19	South Africa	48.8
10	Rwanda	44.6	20	Chad	49.3

Lowest female life expectancy
Years, 2005–10

1	Swaziland	39.4	10	Liberia	46.6
2	Lesotho	42.3	11	Nigeria	47.3
3	Mozambique	42.4	12	Congo-Kinshasa	47.7
4	Zambia	42.5	13	Rwanda	47.8
5	Zimbabwe	42.7	14	Guinea-Bissau	47.9
6	Afghanistan	43.8	15	Malawi	48.4
7	Sierra Leone	44.2	16	Côte d'Ivoire	49.3
8	Angola	44.3	17	Somalia	49.4
9	Central African Rep	46.1	18	South Africa	49.7

Death rates and infant mortality

Highest death rates
Number of deaths per 1,000 population, 2005–10

1	Sierra Leone	22.1	48	Benin	11.2
2	Swaziland	21.2	49	Czech Republic	10.9
3	Angola	20.5	50	Germany	10.7
4	Afghanistan	19.9	51	Portugal	10.6
5	Mozambique	19.8	52	Italy	10.5
6	Lesotho	19.2	53	Gambia, The	10.4
7	Zambia	18.8	54	Denmark	10.3
8	Guinea-Bissau	18.4	55	Kazakhstan	10.1
9	Liberia	18.3		Sudan	10.1
10	Central African Rep	18.1		Sweden	10.1
	Congo-Kinshasa	18.1		Togo	10.1
12	Zimbabwe	17.9	59	Belgium	10.0
13	Rwanda	17.2		Poland	10.0
14	South Africa	17.0		Slovakia	10.0
15	Nigeria	16.8	62	Greece	9.9
16	Somalia	16.6		North Korea	9.9
17	Ukraine	16.4		Slovenia	9.9
18	Russia	16.2		United Kingdom	9.9
19	Burundi	15.6	66	Finland	9.7
20	Chad	15.4		Madagascar	9.7
	Côte d'Ivoire	15.4		Myanmar	9.7
22	Bulgaria	14.8	69	Armenia	9.6
	Equatorial Guinea	14.8		Montenegro	9.6
	Malawi	14.8		Papua New Guinea	9.6
25	Belarus	14.7	72	Bosnia	9.5
	Mali	14.7		Channel Islands	9.5
27	Burkina Faso	14.4	74	Austria	9.4
	Cameroon	14.4	75	Ghana	9.3
29	Estonia	14.3	76	Eritrea	9.2
30	Botswana	14.1		Haiti	9.2
31	Niger	13.8		Macedonia	9.2
32	Latvia	13.6		Uruguay	9.2
33	Uganda	13.4	80	Iraq	9.1
34	Hungary	13.2		Norway	9.1
35	Ethiopia	13.0	82	Cambodia	9.0
36	Tanzania	12.9		Japan	9.0
37	Moldova	12.5		Senegal	9.0
38	Namibia	12.4	85	France	8.9
	Romania	12.4		Timor-Leste	8.9
40	Lithuania	12.3	87	Spain	8.8
41	Croatia	12.1	88	Luxembourg	8.7
42	Guinea	11.9		Faroe Islands[a]	8.7
43	Georgia	11.8	90	Netherlands	8.6
	Kenya	11.8	91	Thailand	8.5
45	Gabon	11.7	92	India	8.2
46	Serbia	11.6		Turkmenistan	8.2
47	Congo-Brazzaville	11.4		United States	8.2

Note: Both death and, in particular, infant mortality rates can be underestimated in certain countries where not all deaths are officially recorded.   a 2007 estimate.

Highest infant mortality
Number of deaths per 1,000 live births, 2005–10

1	Sierra Leone	160.3	21	Equatorial Guinea	92.3
2	Afghanistan	157.0	22	Malawi	89.4
3	Liberia	132.5	23	Togo	88.6
4	Angola	131.9	24	Cameroon	87.5
5	Mali	128.5	25	Ethiopia	86.9
6	Chad	119.2	26	Iraq	81.5
7	Côte d'Ivoire	116.9	27	Uganda	76.9
8	Somalia	116.3	28	Turkmenistan	74.7
9	Congo-Kinshasa	113.5	29	Gambia, The	74.2
10	Guinea-Bissau	112.7	30	Tanzania	72.6
11	Rwanda	112.4	31	Azerbaijan	72.3
12	Niger	110.8	32	Swaziland	71.0
13	Nigeria	109.5	33	Congo-Brazzaville	70.3
14	Burkina Faso	104.4	34	Pakistan	67.5
15	Guinea	102.5	35	Timor-Leste	66.7
16	Burundi	99.4	36	Myanmar	66.0
17	Benin	98.0	37	Senegal	65.7
18	Central African Rep	96.8	38	Madagascar	65.5
19	Mozambique	95.9	39	Sudan	64.9
20	Zambia	92.7	40	Lesotho	64.6

Lowest death rates
No. deaths per 1,000 pop., 2005–10

1	United Arab Emirates	1.4
2	Kuwait	1.9
3	Qatar	2.4
4	Oman	2.7
5	Brunei	2.8
6	Bahrain	3.2
7	Syria	3.4
8	Saudi Arabia	3.7
	West Bank and Gaza	3.7
10	Belize	3.8
11	Jordan	3.9
12	Costa Rica	4.1
	Libya	4.1
14	Malaysia	4.5
15	Cape Verde	4.7
	Macau	4.7
	Nicaragua	4.7
18	Mexico	4.8
	Philippines	4.8
20	Algeria	4.9
21	Cayman Islands[a]	5.0
	Panama	5.0
23	Ecuador	5.1
	Venezuela	5.1
	Vietnam	5.1

Lowest infant mortality
No. deaths per 1,000 live births, 2005–10

1	Iceland	2.9
2	Singapore	3.0
3	Japan	3.2
	Sweden	3.2
5	Norway	3.3
6	Finland	3.7
	Hong Kong	3.7
8	Czech Republic	3.8
9	Andorra[a]	4.0
10	South Korea	4.1
	Switzerland	4.1
12	Belgium	4.2
	France	4.2
	Spain	4.2
15	Germany	4.3
16	Australia	4.4
	Austria	4.4
	Denmark	4.4
19	Luxembourg	4.5
20	Israel	4.7
	Netherlands	4.7
22	Canada	4.8
	Slovenia	4.8
	United Kingdom	4.8
25	Ireland	4.9

a 2007 estimate.

Death and disease

Diabetes
% of population aged 20–79, 2007

1	United Arab Emirates	19.5
2	Saudi Arabia	16.7
3	Kuwait	14.4
4	Oman	13.1
5	Trinidad & Tobago	11.5
6	Mauritius	11.1
7	Egypt	11.0
8	Malaysia	10.7
	Puerto Rico	10.7
10	Mexico	10.6
	Syria	10.6
12	Jamaica	10.3
13	Nicaragua	10.1
	Singapore	10.1
15	Jordan	9.8
16	Panama	9.7
17	Pakistan	9.6
18	Costa Rica	9.3
	Cuba	9.3

Cardiovascular disease
Deaths per 100,000 population, age standardised, 2002

1	Turkmenistan	844
2	Tajikistan	753
3	Kazakhstan	713
4	Afghanistan	706
5	Russia	688
6	Uzbekistan	663
7	Ukraine	637
8	Moldova	619
9	Azerbaijan	613
10	Kyrgyzstan	602
11	Belarus	592
12	Georgia	584
13	Somalia	580
14	Egypt	560
15	Bulgaria	554
16	Yemen	553
17	Turkey	542
18	Albania	537
19	Sierra Leone	515

Cancer
Deaths per 100,000 population, age standardised, 2002

1	Mongolia	306
2	Bolivia	256
3	Hungary	201
4	Sierra Leone	181
5	Poland	180
6	Angola	179
7	Czech Republic	177
8	Peru	175
9	Slovakia	170
	Uruguay	170
11	Liberia	169
	Niger	169
	South Korea	169
14	Croatia	167
	Denmark	167
	Kazakhstan	167
17	Mali	166
18	Burkina Faso	162
	Swaziland	162
20	Congo-Kinshasa	161
	Lithuania	161
22	Côte d'Ivoire	160
	Slovenia	160

Tuberculosis
Incidence per 100,000 population, 2006

1	Swaziland	1,155
2	South Africa	940
3	Namibia	767
4	Lesotho	635
5	Zimbabwe	557
6	Timor-Leste	556
7	Zambia	553
8	Botswana	551
9	Sierra Leone	517
10	Cambodia	500
11	Mozambique	443
12	Côte d'Ivoire	420
13	Congo-Brazzaville	403
14	Rwanda	397
15	Congo-Kinshasa	392
16	Togo	389
17	Kenya	384
18	Ethiopia	378
19	Malawi	377
20	Burundi	367
21	Uganda	355
22	Gabon	354
23	Central African Rep	345

Note: Statistics are not available for all countries. The number of cases diagnosed and reported depends on the quality of medical practice and administration and can be under-reported in a number of countries.

Measles immunisation

*Lowest % of children aged
12–23 months, 2006*

1	Chad	23
2	Central African Rep	35
3	Somalia	35
4	Niger	47
5	Angola	48
	Laos	48
7	Gabon	55
	Venezuela	55
9	Swaziland	57
10	Haiti	58
11	India	59
	Madagascar	59
13	Guinea-Bissau	60
14	Mauritania	62
	Nigeria	62
16	Ethiopia	63
	Namibia	63
18	Timor-Leste	64
19	Papua New Guinea	65
20	Congo-Brazzaville	66

DPTª immunisation

*Lowest % of children aged
12–23 months, 2006*

1	Chad	20
2	Somalia	35
3	Gabon	38
4	Niger	39
5	Central African Rep	40
6	Angola	44
7	Haiti	53
8	Nigeria	54
9	India	55
10	Laos	57
11	Madagascar	61
12	Sierra Leone	64
13	Timor-Leste	67
14	Mauritania	68
	Swaziland	68
16	Indonesia	70
17	Guinea	71
	Venezuela	71
19	Ethiopia	72
	Mozambique	72

HIV/AIDS

*Prevalence among population
aged 15 and over, %, 2005*

1	Swaziland	34.5
2	Botswana	23.6
3	Lesotho	22.7
4	Zimbabwe	19.2
5	Namibia	17.7
6	South Africa	16.6
7	Zambia	15.8
8	Mozambique	14.4
9	Malawi	12.5
10	Central African Rep	10.0
11	Gabon	6.8
12	Côte d'Ivoire	6.4
13	Uganda	6.3
14	Kenya	6.1
15	Tanzania	5.9
16	Cameroon	4.9
17	Congo-Brazzaville	4.7
18	Guinea-Bissau	3.5
	Nigeria	3.5
20	Haiti	3.4
21	Angola	3.3
22	Burundi	3.1
	Chad	3.1
	Rwanda	3.1

AIDS

*Estimated deaths per 100,000
pop., 2005*

1	Swaziland	1,550
2	Zimbabwe	1,384
3	Lesotho	1,282
4	Botswana	1,020
5	Zambia	840
6	Namibia	837
7	Mozambique	707
8	South Africa	675
9	Malawi	605
10	Central African Rep	594
11	Kenya	409
12	Tanzania	365
13	Côte d'Ivoire	358
14	Gabon	340
15	Uganda	316
16	Cameroon	282
17	Congo-Brazzaville	275
18	Rwanda	232
19	Bahamas	200
	Barbados	200
	Belize	200
	Equatorial Guinea	200
	Suriname	200

a Diptheria, pertussis and tetanus

Health

Highest health spending
As % of GDP

1	United States	15.9
2	Timor-Leste	13.7
3	Malawi	12.2
4	Switzerland	11.4
5	France	11.1
6	Germany	10.7
7	Jordan	10.5
8	Argentina	10.2
	Austria	10.2
	Portugal	10.2
11	Greece	10.1
12	Canada	9.7
13	Belgium	9.6
14	Iceland	9.5
15	Lesotho	9.4
16	Netherlands	9.2
17	Denmark	9.1
18	Norway	9.0
19	Italy	8.9
	New Zealand	8.9
	Sweden	8.9
22	Australia	8.8
	Bosnia	8.8
24	Lebanon	8.7
	South Africa	8.7
26	Georgia	8.6
27	Slovenia	8.5
28	Croatia	8.4
	Malta	8.4
30	Nicaragua	8.3

Lowest health spending
As % of GDP

1	Equatorial Guinea	1.7
2	Angola	1.8
3	Congo-Brazzaville	1.9
4	Brunei	2.0
5	Indonesia	2.1
	Pakistan	2.1
7	Kuwait	2.2
	Myanmar	2.2
9	Oman	2.5
10	Somalia	2.6
	United Arab Emirates	2.6
12	Mauritania	2.7
13	Bangladesh	2.8
14	Libya	3.2
	Madagascar	3.2
	Philippines	3.2
17	Burundi	3.4
	Saudi Arabia	3.4
19	Algeria	3.5
	North Korea	3.5
	Singapore	3.5
	Thailand	3.5
23	Laos	3.6
24	Chad	3.7
	Eritrea	3.7
	Sierra Leone	3.7
27	Bahrain	3.8
	Niger	3.8
	Sudan	3.8

Highest pop. per doctor

1	Malawi	49,624
2	Niger	48,649
3	Tanzania	47,445
4	Bhutan	42,308
5	Ethiopia	40,961
6	Mozambique	39,300
7	Burundi	39,000
8	Sierra Leone	35,185
9	Liberia	33,010
10	Chad	28,986
11	Togo	28,000
12	Benin	27,974
13	Botswana	24,658
14	Eritrea	21,395
15	Rwanda	21,296
16	Lesotho	20,225
17	Senegal	20,034
18	Burkina Faso	19,209

Lowest pop. per doctor

1	Turkmenistan	153
2	Cuba	170
3	Greece	200
4	Belarus	209
5	Georgia	214
6	Russia	232
7	Belgium	236
8	Lithuania	252
9	Switzerland	253
10	Malta	256
11	Kazakhstan	263
	Norway	263
13	Israel	265
14	Iceland	267
15	Armenia	270
	Italy	270
17	Netherlands	271
18	Austria	273

Most hospital beds

Beds per 1,000 pop.

1	Japan	14.3		21	Slovakia	6.9
2	North Korea	13.2		22	Romania	6.6
3	Belarus	11.1		23	Bulgaria	6.4
4	Russia	9.7			Moldova	6.4
5	Ukraine	8.7		25	Israel	6.3
6	Czech Republic	8.4			Luxembourg	6.3
	Germany	8.4		27	Tajikistan	6.2
8	Azerbaijan	8.2		28	New Zealand	6.0
9	Lithuania	8.1		29	Serbia	5.9
10	Hungary	7.9		30	Estonia	5.8
11	Austria	7.7		31	Ireland	5.7
	Kazakhstan	7.7			Switzerland	5.7
	Latvia	7.7		33	Croatia	5.5
14	France	7.5		34	Belgium	5.3
	Iceland	7.5			Poland	5.3
	Malta	7.5		36	Uzbekistan	5.2
	Mongolia	7.5		37	Kyrgyzstan	5.1
18	Barbados	7.3		38	Netherlands	5.0
19	South Korea	7.1		39	Cuba	4.9
20	Finland	7.0			Turkmenistan	4.9

Obesity[a]

Men, % of total population

			Women, % of total population		
1	Lebanon	36.3	1 Qatar	45.3
2	Qatar	34.6	2 Saudi Arabia	44.0
3	United States	31.1	3 Lebanon	38.3
4	Panama	27.9	4 Panama	36.1
5	Kuwait	27.5	5 Albania	35.6
6	Cyprus	26.6	6 Mexico	34.5
7	Saudi Arabia	26.4	7 Bahrain	34.1
8	England	24.9	8 Egypt	33.2
9	Mexico	24.4	United States	33.2
10	Austria	23.3	10 United Arab Emirates	31.4
	Bahrain	23.3	11 Kuwait	29.9
12	Canada	22.9	12 Turkey	29.4
	Malta	22.9	13 Scotland	26.0
14	Albania	22.8	14 England	25.2
15	Germany	22.5	15 Mongolia	24.6
16	Scotland	22.4	16 Jamaica	23.9
17	New Zealand	21.2	17 Oman	23.8
18	Greece	20.0	18 Cyprus	23.7
19	Argentina	19.5	19 Germany	23.3
20	Australia	19.3	20 Canada	23.2
21	Luxembourg	18.8	21 Chile	23.0
22	Wales	18.0	Peru	23.0
23	United Arab Emirates	17.1	23 Australia	22.2
24	Oman	16.7	24 New Zealand	22.1
			25 Morocco	21.7

Note: Data for these health rankings refer to the latest year available, 1999–2006
a Defined as body mass index of 30 or more – see page 248.

Teenagers' health behaviour[a]

Computer use

% using a computer two or more hours every weekday

1	Netherlands	68	13	Finland	47
2	Iceland	60		Portugal	47
3	England	57	15	Luxembourg	46
	Estonia	57	16	Denmark	45
	Israel	57		Germany	45
6	Norway	55	18	Bulgaria	44
	Sweden	55		Latvia	44
8	Canada	54	20	Belgium (French)	39
	Wales	54		Slovakia	39
10	Belgium (Flemish)	52	22	Austria	38
11	Poland	51		Turkey	38
	Scotland	51		United States	38

Obesity[b]

% overweight or obese

Girls			Boys		
1	Malta	28	1	United States	33
2	United States	26	2	Malta	32
3	Greenland	23	3	Canada	25
4	Wales	18		Greece	25
5	Canada	14	5	Italy	23
6	Portugal	13	6	Iceland	22
7	Belgium (French)	12		Portugal	22
	Finland	12	8	Greenland	21
	Iceland	12		Wales	21
	Scotland	12	10	Slovenia	20
11	Germany	11	11	Austria	19
	Greece	11		Croatia	19
	Hungary	11		Finland	19
	Spain	11		Macedonia	19
15	Croatia	10		Spain	19
	Ireland	10	16	Bulgaria	18
	Italy	10	17	Hungary	17
	Netherlands	10		Israel	17
	Slovenia	10			

Smoking

% smoking every day

1	Greenland	34		Estonia	16
2	Bulgaria	26		Lithuania	16
3	Croatia	20	13	Germany	15
4	Austria	19		Ireland	15
	Latvia	19		Scotland	15
	Ukraine	19	16	France	14
7	Hungary	18		Italy	14
	Russia	18		Luxembourg	14
9	Finland	17		Netherlands	14
10	Czech Republic	16			

a WHO survey covering 41 countries in Europe and North America. Data is for 15 year-olds.
b According to body-mass index (see Glossary page 248).

Cannabis

% using in past 12 months

Regular user (3–39 times)

1	Canada	14
2	United States	12
3	Spain	11
	Wales	11
5	Switzerland	10
6	France	9
	Netherlands	9
	Scotland	9
9	Belgium (Flemish)	8
	Czech Republic	8
	England	8
	Ireland	8
	Italy	8
	Luxembourg	8
15	Belgium (French)	7
	Estonia	7

Heavy user (40 times or more)

1	Canada	5
2	Spain	4
	Switzerland	4
4	Belgium (French)	3
	France	3
	Ireland	3
	Luxembourg	3
	Netherlands	3
	Scotland	3
	United States	3
	Wales	3
12	Croatia	2
	Czech Republic	2
	England	2
	Greenland	2

Alcohol

% drinking at least once a week

Beer

1	Ukraine	44
2	Czech Republic	28
3	Bulgaria	27
4	Italy	24
	Netherlands	24
	Denmark	24
7	England	23
	Croatia	23

Wine

1	Malta	23
2	Croatia	18
3	Italy	15
4	Hungary	13
	Slovenia	13
6	Austria	12

Spirits

1	Malta	26
2	Austria	20
3	Denmark	17
	Scotland	17
	Spain	17
6	Greece	15

Alcopops

1	Ukraine	22
2	Austria	21
3	England	19
	Wales	19
5	Malta	18
	Netherlands	18
7	Belgium (French)	17
	Scotland	17

Fighting

% involved in a physical fight at least once in the past 12 months

1	Malta	49		Macedonia	40
2	Belgium (French)	46	10	Lithuania	39
3	Greece	44		Ukraine	39
	Turkey	44	12	Netherlands	38
5	Slovakia	43	13	England	37
6	Czech Republic	41		Romania	37
	Ireland	41		Russia	37
8	Austria	40		Scotland	37

Marriage and divorce

Highest marriage rates

Number of marriages per 1,000 population, 2006 or latest available year

1	Cayman Islands	16.8	31	South Korea	6.6
2	Bermuda	13.7		Turkmenistan	6.6
3	Vietnam	12.1	33	Turkey	6.4
4	Iran	11.5	34	China	6.3
5	Jordan	9.9	35	Romania	6.2
6	Mauritius	9.1	36	Belize	6.1
7	Fiji	9.0	37	Costa Rica	6.0
8	Algeria	8.7		Hong Kong	6.0
9	Azerbaijan	8.5		Lithuania	6.0
10	Guam	8.2		Puerto Rico	6.0
	Jamaica	8.2	41	Malaysia	5.9
12	Lebanon	8.0		Malta	5.9
13	United States	7.9		Mongolia	5.9
14	Cyprus	7.8		Tunisia	5.9
15	West Bank and Gaza	7.7		Ukraine	5.9
16	Channel Islands[a]	7.6	46	Bahamas	5.8
	Russia	7.6		Trinidad & Tobago	5.8
	Tajikistan	7.6	48	Aruba	5.7
19	Belarus	7.4		Finland	5.7
	Taiwan	7.4		Mexico	5.7
21	Kyrgyzstan	7.3	51	Bosnia	5.6
22	Indonesia	7.2		Greece	5.6
23	Denmark	7.1	53	Australia	5.5
	Macedonia	7.1		Japan	5.5
25	Kazakhstan	7.0	55	Iceland	5.4
	Moldova	7.0	56	Spain	5.3
27	Albania	6.9		Switzerland	5.3
28	Uzbekistan	6.8	58	Armenia	5.2
29	Egypt	6.7		Singapore	5.2
	Philippines	6.7		United Kingdom	5.2

Lowest marriage rates

Number of marriages per 1,000 population, 2006 or latest available year

1	Colombia	1.7	12	Macau	3.7
2	Dominican Republic	2.6		Martinique	3.7
3	Venezuela	2.7	14	El Salvador	3.8
4	Andorra	2.8		South Africa	3.8
	Peru	2.8	16	Guadeloupe	3.9
6	Argentina	3.2		Netherlands Antilles	3.9
	Slovenia	3.2		New Caledonia	3.9
	United Arab Emirates	3.2	19	Bulgaria	4.0
9	Chile	3.3		Réunion	4.0
	Panama	3.3		Suriname	4.0
11	Qatar	3.4		Uruguay	4.0

a Jersey only
Note: The data are based on latest available figures and hence will be affected by the population age structure at the time. Marriage rates refer to registered marriages only and, therefore, reflect the customs surrounding registry and efficiency of administration.

Highest divorce rates
Number of divorces per 1,000 population, 2006 or latest available year

1	Aruba	4.4		Norway	2.5
2	Uruguay	4.3	27	Hong Kong	2.4
3	Russia	4.1		Hungary	2.4
	South Korea	4.1		Kazakhstan	2.4
5	Moldova	4.0		Portugal	2.4
6	Puerto Rico	3.9	31	Austria	2.3
7	Ukraine	3.6		Costa Rica	2.3
	United States	3.6		Latvia	2.3
9	Lithuania	3.4		Luxembourg	2.3
10	Czech Republic	3.3	35	Bulgaria	2.2
	Taiwan	3.3		Canada	2.2
12	Channel Islands[a]	3.2		France	2.2
13	Cuba	3.1		Slovakia	2.2
	Estonia	3.1		Sweden	2.2
15	Belgium	3.0	40	Guadeloupe	2.1
	Denmark	3.0	41	Cyprus	2.0
17	Bermuda	2.9		Japan	2.0
	United Kingdom	2.9	43	Iceland	1.9
19	Belarus	2.8		Jordan	1.9
	Germany	2.8	45	Kuwait	1.8
	Netherlands Antilles	2.8		Netherlands	1.8
22	New Zealand	2.7	47	Romania	1.7
23	Switzerland	2.6		Spain	1.7
24	Australia	2.5	49	Israel	1.6
	Finland	2.5		Singapore	1.6

Lowest divorce rates
Number of divorces per 1,000 population, 2006 or latest available year

1	Belize	0.2		Ireland	0.8
	Colombia	0.2		Italy	0.8
3	Libya	0.3		Macedonia	0.8
4	Georgia	0.4		Turkey	0.8
	Tajikistan	0.4		Venezuela	0.8
6	Bosnia	0.5	23	Ecuador	0.9
	Chile	0.5		Mauritius	0.9
	Vietnam	0.5		Panama	0.9
9	El Salvador	0.6		Saudi Arabia	0.9
	Mongolia	0.6		United Arab Emirates	0.9
	Uzbekistan	0.6	28	South Africa	1.0
12	Jamaica	0.7	29	Azerbaijan	1.1
	Mexico	0.7		New Caledonia	1.1
14	Armenia	0.8		Réunion	1.1
	Brazil	0.8		Thailand	1.1
	Egypt	0.8		Tunisia	1.1
	Indonesia	0.8		West Bank and Gaza	1.1

a Jersey only

Households and living costs

Number of households[a]

Biggest, m, 2006

1	China	376.5
2	India	209.9
3	United States	113.9
4	Indonesia	59.6
5	Russia	53.0
6	Brazil	50.7
7	Japan	48.8
8	Germany	39.4
9	Nigeria	27.5
10	United Kingdom	26.2
11	Bangladesh	25.7
12	Vietnam	25.6
13	France	25.5
14	Mexico	25.1
15	Pakistan	22.7
16	Italy	22.6
17	Ukraine	19.8
18	South Korea	17.6
19	Congo-Kinshasa	17.5
	Philippines	17.5
	Thailand	17.5
22	Egypt	16.4
23	Turkey	15.7
24	Spain	15.3

Smallest, m, 2006

1	Cayman Islands	0.01
2	Bermuda	0.02
3	Aruba	0.03
	Guam	0.03
5	French Polynesia	0.04
	New Caledonia	0.04
7	Netherlands Antilles	0.05
8	Barbados	0.06
9	Bahamas	0.07
	Brunei	0.07
11	Suriname	0.09
12	Martinique	0.10
13	Belize	0.11
	Guadeloupe	0.11
15	Cape Verde	0.12
	Malta	0.12
17	Equatorial Guinea	0.13
	Iceland	0.13
	Réunion	0.13
20	Macau	0.15
21	Fiji	0.16
	Qatar	0.16
23	Luxembourg	0.18
24	Bahrain	0.19

Size of households[a]

Population per dwelling, 2006

Biggest

1	Congo-Brazzaville	8.2
2	Pakistan	7.1
3	Papua New Guinea	6.9
4	United Arab Emirates	6.6
5	Cambodia	6.3
	Guinea	6.3
	Kuwait	6.3
8	Gabon	6.2
9	Réunion	6.1
	Sudan	6.1
11	Bangladesh	6.0
	Uzbekistan	6.0
13	Algeria	5.9
	Kyrgyzstan	5.9
15	French Polynesia	5.8
	Saudi Arabia	5.8
17	Burundi	5.7
	Guinea-Bissau	5.7
19	Jordan	5.6

Smallest

1	Germany	2.1
	Sweden	2.1
3	Denmark	2.2
	Finland	2.2
	Switzerland	2.2
6	Iceland	2.3
	Netherlands	2.3
	Norway	2.3
	Ukraine	2.3
	United Kingdom	2.3
11	Austria	2.4
	Belgium	2.4
	Estonia	2.4
	France	2.4
15	Azerbaijan	2.5
	Belarus	2.5
	Lithuania	2.5
	Slovakia	2.5

a Latest available year.

Cost of living[a]

End 2007, USA=100

Highest			Lowest		
1	Norway	148	1	Iran	38
2	France	141	2	Venezuela	38
3	United Kingdom	139	3	Pakistan	44
4	Denmark	137	4	Philippines	46
5	Iceland	128	5	Libya	48
6	Finland	125	6	India	51
	Japan	125		Nepal	51
8	Austria	119	8	Bangladesh	53
9	Switzerland	117	9	Paraguay	54
10	Australia	116	10	Algeria	57
11	Germany	115		Argentina	57
	Ireland	115		Costa Rica	57
	South Korea	115	13	Cambodia	58
14	Belgium	114		Sri Lanka	58
15	New Caledonia	113	15	Syria	59
16	Singapore	112		Uzbekistan	59
	Sweden	112			
18	Netherlands	108			
	Spain	108			

Mortgage debt

As % of GDP, 2006

1	Switzerland	132.3	18	France	31.8
2	Iceland[b]	102.1	19	Latvia	28.9
3	Denmark	98.1	20	Greece	26.8
4	United Kingdom	82.8	21	Austria	23.6
5	United States	79.0	22	Italy	16.6
6	Netherlands	72.6	23	Cyprus[b]	16.0
7	Ireland	63.4	24	Croatia[b]	12.7
8	Portugal	59.2	25	Lithuania	12.6
9	Spain	56.1	26	Czech Republic	12.0
10	Norway[b]	55.1	27	Hungary	11.9
11	Germany	42.3	28	Slovakia	10.4
12	Sweden	41.3	29	Poland	7.5
13	Luxembourg	36.4	30	Slovenia	6.6
14	Malta	34.8	31	Bulgaria[b]	4.7
15	Belgium	34.2	32	Turkey[b]	2.5
16	Finland	32.9	33	Romania	2.3
17	Estonia	32.4			

a The cost of living index shown is compiled by the Economist Intelligence Unit for use by companies in determining expatriate compensation: it is a comparison of the cost of maintaining a typical international lifestyle in the country rather than a comparison of the purchasing power of a citizen of the country. The index is based on typical urban prices an international executive and family will face abroad. The prices are for products of international comparable quality found in a supermarket or department store. Prices found in local markets and bazaars are not used unless the available merchandise is of the specified quality and the shopping area itself is safe for executive and family members. New York City prices are used as the base, so United States = 100.

b 2005

Consumer goods ownership

TV
Colour TVs per 100 households, 2006

1	United Arab Emirates	99.7		Portugal	98.7
2	Taiwan	99.6	18	China	98.5
3	Hong Kong	99.5		Poland	98.5
	Ireland	99.5		United States	98.5
	Japan	99.5	21	Netherlands	98.3
	United Kingdom	99.5		Switzerland	98.3
7	Greece	99.4	23	Norway	98.2
	Singapore	99.4	24	Denmark	98.0
	Spain	99.4		New Zealand	98.0
10	South Korea	99.2	26	Germany	97.7
11	Kuwait	99.1	27	Czech Republic	97.5
12	Canada	99.0	28	Sweden	97.3
13	Australia	98.8	29	Saudi Arabia	97.2
	Belgium	98.8	30	Finland	96.9
15	Austria	98.7	31	Slovenia	96.7
	Hungary	98.7	32	Italy	96.6

Telephone
Telephone lines per 100 people, 2006

1	Bermuda	89.5	18	Andorra	51.3
2	Switzerland	66.9	19	Barbados	50.1
3	Germany	65.9	20	Malta	50.0
4	Canada	64.5	21	Ireland	49.9
5	Taiwan	63.6	22	South Korea	49.8
6	Iceland	63.5	23	Australia	48.8
	Virgin Islands (US)	63.5	24	Faroe Islands	48.7
8	Greenland	62.8	25	Cyprus	48.3
9	Sweden	59.5	26	Netherlands	46.6
10	Montenegro	58.9	27	Italy	46.3
11	United States	57.2	28	Spain	45.8
12	Denmark	56.9	29	Belgium	45.2
13	United Kingdom	56.2	30	Norway	44.3
14	France	55.8	31	New Zealand	44.1
15	Greece	55.4	32	Israel	43.9
16	Hong Kong	53.9	33	Austria	43.4
17	Luxembourg	52.4	34	Japan	43.0

CD player
CD players per 100 households, 2006

1	New Zealand	88.5	12	Austria	69.5
2	United Kingdom	88.4	13	Belgium	65.5
3	Denmark	88.1	14	Finland	63.8
	Norway	88.1	15	United States	61.3
5	Netherlands	87.0	16	Switzerland	59.2
6	Australia	85.5	17	Hong Kong	58.4
7	Germany	84.1	18	Singapore	56.0
8	Sweden	82.9	19	Portugal	42.6
9	Canada	81.5	20	Spain	42.1
10	Taiwan	70.6	21	Ireland	40.7
11	Japan	69.8	22	Peru	36.3

Computer

Computers per 100 people, 2006

1	Israel	122.1	25	Slovakia	35.8
2	Canada	87.6	26	Spain	27.7
3	Switzerland	86.5	27	Czech Republic	27.4
4	Netherlands	85.4	28	United Arab Emirates	25.6
5	Sweden	83.6	29	Latvia	24.6
6	United States	76.2	30	Poland	24.2
7	United Kingdom	75.8	31	Kuwait	23.7
8	Australia	75.7	32	Costa Rica	23.1
9	Denmark	69.6	33	Macedonia	22.2
10	Singapore	68.2	34	Malaysia	21.8
11	Japan	67.6	35	Croatia	19.9
12	Hong Kong	61.2	36	Lithuania	18.0
13	Austria	60.7	37	Mauritius	16.9
14	Germany	60.6	38	Brazil	16.1
15	Norway	59.4	39	Hungary	14.9
16	France	57.5	40	Chile	14.1
17	South Korea	53.2	41	Mexico	13.6
18	Ireland	52.8		Saudi Arabia	13.6
19	New Zealand	50.2		Uruguay	13.6
20	Finland	50.0	44	Mongolia	13.3
21	Estonia	48.3		Portugal	13.3
22	Slovenia	40.4	46	Romania	12.9
23	Belgium	37.7	47	Namibia	12.3
24	Italy	36.7	48	Russia	12.2

Mobile telephone

Subscribers per 100 people, 2006

1	Lithuania	138.1	25	Spain	106.4
2	Macau	137.4	26	Netherlands	106.0
3	Italy	135.1	27	Sweden	105.9
4	Hong Kong	132.7	28	Faroe Islands	105.7
5	Trinidad & Tobago	126.4		Russia	105.7
6	Estonia	125.2	30	Aruba	104.9
7	Bahrain	122.9	31	Germany	103.6
8	Israel	122.7	32	Cyprus	102.8
9	Czech Republic	121.5	33	Taiwan	102.0
10	United Arab Emirates	118.5	34	Hungary	99.0
11	Luxembourg	116.8		Switzerland	99.0
12	United Kingdom	116.6	36	Greece	98.6
13	Portugal	116.0	37	Australia	97.0
14	Austria	112.8	38	Andorra	96.9
15	Ireland	112.6	39	Croatia	96.5
16	Qatar	109.6	40	Poland	95.5
17	Singapore	109.3	41	Latvia	95.1
18	Iceland	108.7	42	Greenland	94.1
19	Norway	108.6	43	New Zealand	94.0
20	Finland	107.8	44	Jamaica	93.7
21	Bulgaria	107.6	45	Bermuda	93.3
22	Montenegro	107.3	46	Belgium	92.6
23	Denmark	107.0		Slovenia	92.6
24	Ukraine	106.5	48	Kuwait	91.5

Cinema and films

Cinema attendances

Total visits, m, 2006		Visits per head, 2006	
1 India	1,473.4	1 New Zealand	8.1
2 China	1,457.9	2 Australia	6.6
3 United States	1,447.7	3 United States	4.8
4 Indonesia	284.1	4 Iceland	4.7
5 France	188.7	5 Ireland	4.3
6 Mexico	169.2	6 Canada	4.1
7 Japan	164.3	7 Singapore	3.5
8 United Kingdom	156.6	8 France	3.1
9 Germany	136.7	9 Malta	2.8
10 Australia	135.6	Spain	2.8
11 Canada	135.1	11 Norway	2.6
12 Spain	121.7	United Kingdom	2.6
13 Italy	107.3	13 Luxembourg	2.5
14 Philippines	92.4	14 Belgium	2.3
15 Russia	89.8	Denmark	2.3
16 South Africa	65.8	Venezuela	2.3
17 Venezuela	63.7	17 Switzerland	2.2
18 Argentina	47.8	18 Austria	2.1
19 South Korea	47.5	19 Italy	1.8
20 Brazil	44.8	20 Germany	1.7
21 Turkey	34.9	Sweden	1.7
22 New Zealand	33.4	22 Mexico	1.6
23 Poland	32.0	Portugal	1.6
24 Belgium	23.8	24 Ecuador	1.5
25 Netherlands	22.5	25 Netherlands	1.4
26 Ecuador	20.1	Slovenia	1.4
27 Taiwan	18.5	South Africa	1.4
28 Ireland	17.9	28 Finland	1.3
29 Austria	17.3	India	1.3
30 Portugal	16.4	Indonesia	1.3
Switzerland	16.4	Israel	1.3
32 Malaysia	16.3	Japan	1.3

Top Oscar winners

	Film	Awards	Nominations
1	Ben-Hur (1959)	11	12
	Titanic (1997)	11	14
	The Lord of the Rings: The Return of the King (2003)	11	11
4	West Side Story (1961)	10	11
5	Gigi (1958)	9	9
	The Last Emperor (1987)	9	9
	The English Patient (1996)	9	12
8	Gone with the Wind (1939)	8	13
	From Here to Eternity (1953)	8	13
	On the Waterfront (1954)	8	12
	My Fair Lady (1964)	8	12
	Cabaret[a] (1972)	8	10
	Gandhi (1982)	8	11
	Amadeus (1984)	8	11

a Did not win best picture award.

The press

Daily newspapers

Copies per '000 population, latest year

1	Japan	551	16	New Zealand	185
2	Norway	517	17	Czech Republic	182
3	Sweden	480		Ireland	182
4	Finland	431	19	Canada	175
5	Switzerland	429		Slovenia	175
6	Singapore	361	21	France	165
7	Denmark	352	22	Belgium	164
8	Austria	315	23	Australia	156
9	Netherlands	308	24	Latvia	154
10	United Kingdom	292	25	Trinidad & Tobago	151
11	Germany	267	26	Spain	145
12	Hong Kong	223	27	Italy	138
13	Hungary	217	28	Ukraine	132
14	United States	194	29	Slovakia	125
15	Estonia	192	30	Poland	113

Press freedom[a]

Scores, 2007

Most free			Least free		
1	Iceland	0.75	1	Eritrea	114.75
	Norway	0.75	2	North Korea	108.75
3	Estonia	1.00	3	Turkmenistan	103.75
	Slovakia	1.00	4	Iran	96.50
5	Belgium	1.50	5	Cuba	96.17
	Finland	1.50	6	Myanmar	93.75
	Sweden	1.50	7	China	89.00
8	Denmark	2.00	8	Vietnam	79.25
	Ireland	2.00	9	Laos	75.00
	Portugal	2.00	10	Uzbekistan	74.88
11	Switzerland	3.00	11	Somalia	71.50
12	Latvia	3.50	12	West Bank and Gaza	69.83
	Netherlands	3.50	13	Iraq	67.83
14	Czech Republic	4.00	14	Sri Lanka	67.50
15	New Zealand	4.17	15	Libya	66.50
16	Austria	4.25	16	Syria	66.00
17	Hungary	4.50	17	Equatorial Guinea	65.25
18	Canada	4.88	18	Pakistan	64.83
19	Trinidad & Tobago	5.00	19	Belarus	63.63
20	Germany	5.75	20	Ethiopia	63.00
21	Costa Rica	6.50	21	Zimbabwe	62.00
	Slovenia	6.50	22	Saudi Arabia	59.75
23	Lithuania	7.00	23	Rwanda	58.88
24	United Kingdom	8.25	24	Egypt	58.00
25	Mauritius	8.50	25	Tunisia	57.00
	Namibia	8.50	26	Russia	56.90
27	Jamaica	8.63	27	Yemen	56.67
28	Australia	8.79	28	Afghanistan	56.50
29	Ghana	9.00	29	Singapore	56.00
30	Greece	9.25	30	Sudan	55.75

a Based on 50 questions about press freedom.

Music and the internet

Music sales

	Total including downloads, $m, 2006			$ per head, 2006	
1	United States	6,497	1	United Kingdom	34.3
2	Japan	3,563	2	Japan	27.8
3	United Kingdom	2,054	3	Norway	26.1
4	Germany	1,411	4	Switzerland	24.9
5	France	1,126	5	United States	21.6
6	Canada	530	6	Australia	19.8
7	Australia	403	7	France	18.6
8	Italy	383	8	Germany	17.1
9	Spain	327	9	Canada	16.3
10	Mexico	236	10	Austria	15.9
11	Netherlands	233	11	Belgium	15.6
12	Brazil	222	12	Sweden	15.5
13	Russia	210	13	Netherlands	14.2
14	Switzerland	182	14	Spain	7.5
15	Belgium	162	15	Italy	6.6
16	South Africa	154	16	South Africa	3.2
17	South Korea	153		South Korea	3.2
18	Sweden	141	18	Mexico	2.2
19	Austria	130	19	Russia	1.5
20	Norway	120	20	Brazil	1.2

Internet hosts

	By country, January 2008			Per 1,000 pop., January 2008	
1	United States[a]	291,214,772	1	United States[a]	967.5
2	Japan	36,803,719	2	Iceland	768.9
3	Germany	20,659,105	3	Finland	703.5
4	Italy	16,730,591	4	Netherlands	642.7
5	France	14,356,747	5	Denmark	603.0
6	China	13,113,985	6	Norway	592.4
7	Australia	10,707,139	7	Australia	524.9
8	Netherlands	10,540,083	8	Switzerland	453.2
9	Brazil	10,151,592	9	Estonia	434.3
10	Mexico	10,071,370	10	New Zealand	411.6
11	United Kingdom	7,727,550	11	Sweden	386.1
12	Poland	7,134,976	12	Belgium	347.9
13	Taiwan	5,121,607	13	Luxembourg	334.8
14	Canada	4,717,308	14	Andorra	328.4
15	Finland	3,728,551	15	Austria	315.8
16	Belgium	3,618,495	16	Ireland	297.1
17	Russia	3,577,635	17	Italy	288.0
18	Sweden	3,513,170	18	Japan	287.1
19	Switzerland	3,308,684	19	Greenland	252.3
20	Denmark	3,256,134	20	Germany	249.8
21	Argentina	3,128,975	21	France	236.5
22	Spain	3,085,513	22	Lithuania	230.5
23	Norway	2,725,031	23	Taiwan	224.6
24	Austria	2,589,316	24	Netherlands Antilles	215.0
25	India	2,584,572	25	Israel	205.6
26	Turkey	2,425,789	26	Czech Republic	205.2

a Includes all hosts ending ".com", ".net" and ".org", which exaggerates the numbers.

Nobel prize winners: 1901–2007

Peace (two or more)

1	United States	18
2	United Kingdom	11
3	France	9
4	Sweden	5
5	Belgium	4
	Germany	4
7	Austria	3
	Norway	3
	South Africa	3
	Switzerland	3
11	Argentina	2
	Egypt	2
	Israel	2
	Russia	2

Economics[a]

1	United States	31
2	United Kingdom	8
3	Norway	2
	Sweden	2
5	France	1
	Germany	1
	Israel	1
	Netherlands	1
	Russia	1

Literature (three or more)

1	France	14
2	United States	12
3	United Kingdom	11
4	Germany	7
5	Sweden	6
6	Italy	5
	Spain	5
8	Norway	3
	Poland	3
	Russia	3

Medicine (three or more)

1	United States	51
2	United Kingdom	21
3	Germany	14
4	Sweden	7
5	France	6
	Switzerland	6
7	Austria	5
	Denmark	5
9	Australia	3
	Belgium	3
	Italy	3

Physics

1	United States	48
2	United Kingdom	19
3	Germany	19
4	France	9
5	Netherlands	6
	Russia	6
7	Japan	4
	Sweden	4
	Switzerland	4
10	Austria	3
	Italy	3
12	Canada	2
	Denmark	2
14	Colombia	1
	India	1
	Ireland	1
	Pakistan	1
	Poland	1

Chemistry

1	United States	42
2	United Kingdom	22
3	Germany	15
4	France	7
5	Switzerland	6
6	Sweden	5
7	Canada	4
	Japan	4
9	Argentina	1
	Austria	1
	Belgium	1
	Czech Republic	1
	Denmark	1
	Finland	1
	Israel	1
	Italy	1
	Netherlands	1
	Norway	1
	Russia	1

a Since 1969.

Notes: Prizes by country of residence at time awarded. When prizes have been shared in the same field, one credit given to each country. Only top rankings in each field are included.

Drinking and smoking

Beer drinkers

Off-trade sales, litres per head of pop., 2006

1	Czech Republic	81.9
2	Venezuela	76.7
3	Australia	69.3
4	Germany	67.9
5	Austria	66.2
6	Finland	65.4
7	Russia	62.6
8	United States	62.5
9	Slovakia	62.4
10	Denmark	60.8
11	Hungary	60.4
12	New Zealand	55.0
13	Poland	54.5
14	Netherlands	54.3
15	Canada	52.7
16	Romania	51.3
17	Bulgaria	45.8
18	Mexico	45.3
19	Ukraine	43.5
20	Belgium	42.8
21	Sweden	41.0
22	Norway	40.3

Wine drinkers

Off-trade sales, litres per head of pop., 2006

1	Portugal	33.1
2	Switzerland	29.0
3	Italy	28.7
4	France	26.4
5	Denmark	26.2
6	Argentina	24.3
7	Hungary	23.7
8	Netherlands	22.6
9	Germany	21.1
10	Belgium	19.8
11	New Zealand	18.2
12	United Kingdom	17.7
13	Australia	17.6
14	Austria	17.2
15	Sweden	16.6
16	Ireland	14.9
17	Chile	13.6
18	Spain	12.7
19	Norway	12.3
20	Greece	11.8
21	Czech Republic	10.4
22	Finland	9.9

Alcoholic drinks

Off-trade sales, litres per head of pop., 2006

1	Australia	100.8
2	Czech Republic	97.7
3	Germany	96.4
4	Finland	93.3
5	Austria	90.2
6	Denmark	89.5
7	Hungary	88.3
8	Russia	86.4
9	New Zealand	81.3
10	Netherlands	80.4
	Venezuela	80.4
12	Slovakia	75.2
13	United States	74.8
14	Poland	69.4
15	Portugal	68.9
16	Canada	68.7
17	United Kingdom	66.0
18	Belgium	65.8
19	Sweden	64.1
20	Argentina	62.3
21	Romania	61.9
22	Switzerland	61.1
23	Norway	56.7

Smokers

Av. ann. consumption of cigarettes per head per day, 2007

1	Greece	8.2
2	Slovenia	7.0
	Ukraine	7.0
4	Bulgaria	6.7
5	Czech Republic	6.5
6	Macedonia	6.4
	Russia	6.4
8	Moldova	6.3
9	Spain	6.1
10	Bosnia	5.8
	Serbia	5.8
12	Armenia	5.7
13	Japan	5.6
14	Belarus	5.4
15	Latvia	5.2
16	Croatia	5.1
	Taiwan	5.1
18	Cyprus	5.0
	Lebanon	5.0
	Poland	5.0
21	Kazakhstan	4.9
22	Estonia	4.7
	South Korea	4.7

Crime and punishment

Murders

Per 100,000 pop., 2004

1	Ecuador	18.9
2	Swaziland	13.6
3	Mongolia	13.1
4	Suriname	10.4
5	Lithuania	9.3
6	Latvia	8.5
	Zimbabwe	8.5
8	Belarus	8.2
	Kyrgyzstan	8.2
10	Turkmenistan	8.1
11	Uganda	7.9
12	Ukraine	7.3
13	Estonia	6.8
	Sri Lanka	6.8
15	Moldova	6.7
16	Costa Rica	6.5
17	Georgia	6.1
18	Peru	5.7
	Uruguay	5.7
20	Philippines	4.5

Death row[a]

Population, latest available year

1	Pakistan	7,436
2	United States	3,246
3	Thailand	1,140
4	Kenya	946
5	Bangladesh	860
6	Burundi	533
7	Uganda	417
8	Tanzania	389
9	Nigeria	341
10	Sudan	300
11	Congo-Kinshasa[b]	200
12	Ghana	152
13	Taiwan[b]	100
14	Indonesia	90
15	South Korea	58
16	Cuba[b]	50
	Iraq[b]	50
18	Guatemala	34
19	Zimbabwe	26
20	Afghanistan[b]	25

Prisoners

Total prison pop., latest available year

1	United States	2,258,983
2	China	1,565,771
3	Russia	885,014
4	Brazil	419,551
5	India	358,368
6	Mexico	217,436
7	South Africa	166,267
8	Thailand	165,316
9	Iran	158,351
10	Ukraine	150,950
11	Indonesia	116,688
12	Vietnam	98,556
13	United Kingdom	91,300
14	Turkey[b]	90,000
15	Philippines	89,639
16	Pakistan	89,370
17	Poland	88,620
18	Bangladesh[b]	86,000
19	Rwanda[b]	82,000
20	Japan[b]	81,300
21	Germany	72,656
22	Spain	67,783
23	Ethiopia[b]	65,000
24	Colombia	63,603

Per 100,000 pop., latest available year

1	United States	751
2	Russia	627
3	Virgin Islands (US)	549
4	Cuba[b]	531
5	Turkmenistan[b]	489
6	Bahamas	462
7	Belize	460
8	Georgia	428
9	Belarus	426
10	Bermuda	394
11	Cayman Islands	391
12	Barbados	379
13	Kazakhstan	378
14	Netherlands Antilles	364
15	Puerto Rico	356
	Suriname	356
17	South Africa	348
18	Panama	337
19	Botswana	329
20	Ukraine	328
21	Aruba	324
22	Israel	305
23	Trinidad & Tobago	288
	United Arab Emirates	288

a Number of prisoners sentenced to death.   b Estimate.

Stars...

Space missions
Firsts and selected events

1957 Man-made satellite Dog in space, Laika
1961 Human in space, Yuri Gagarin
       Entire day in space, Gherman Titov
1963 Woman in space, Valentina Tereshkova
1964 Space crew, one pilot and two passengers
1965 Space walk, Alexei Leonov
       Eight days in space achieved (needed to travel to moon and back)
1966 Docking between space craft and target vehicle
       Autopilot re-entry and landing
1968 Live television broadcast from space
       Moon orbit
1969 Astronaut transfer from one craft to another in space
       Moon landing
1971 Space station, Salyut
       Drive on the moon
1973 Space laboratory, Skylab
1978 Non-Amercian, non-Soviet, Vladimir Remek (Czechoslovakia)
1982 Space shuttle, Columbia (first craft to carry four crew members)
1983 Five crew mission
1984 Space walk, untethered
       Capture, repair and redeployment of satellite in space
       Seven crew mission
1985 Classified US Defence Department mission
1986 Space shuttle explosion, Challenger
       Mir space station activated
1990 Hubble telescope deployed
2001 Dennis Tito, first paying space tourist
2003 Space shuttle explosion, Columbia. Shuttle programme suspended
       China's first manned space flight, Yang Liwei
2004 SpaceShipOne, first successful private manned space flight
2005 Space shuttle, resumption of flights
2008 *Phoenix* lander, mission on Mars

Space vehicle launches[a]

2004		
1 United States	21	
2 Russia	17	
3 China	2	
France	2	
5 India	1	
Sweden	1	

2006		
1 United States	20	
2 Russia	19	
3 Japan	7	
4 France	5	
5 China	4	
6 Sweden	3	

2005		
1 Russia	21	
2 United States	15	
3 France	5	
4 China	3	
5 Japan	2	
6 India	1	
Sweden	1	

2007		
1 Russia	22	
2 United States	20	
3 China	6	
4 France	5	
5 India	3	
Japan	3	
7 Brazil	2	
8 Iran	1	

a By host country and including
   suborbital launches.

...and Wars

Defence spending
As % of GDP, 2006

1	Myanmar	18.7	16	Qatar	4.5
2	Oman	9.0	17	Georgia	4.4
3	Saudi Arabia	8.5	18	Yemen	4.2
4	Israel	7.9	19	Russia	4.1
5	Jordan	7.9	20	Colombia	4.0
6	United Arab Emirates	6.7		Cuba	4.0
7	Eritrea[a]	6.3		Egypt	4.0
8	Armenia[a]	5.6		Guinea-Bissau	4.0
	Vietnam	5.6		United States	4.0
10	Madagascar	5.4		Zimbabwe[a]	4.0
11	Burundi	5.1	26	Morocco	3.8
	Syria	5.1	27	Bahrain	3.4
13	Singapore	4.8		Kuwait	3.4
14	Angola	4.7		Sri Lanka	3.4
15	Uzbekistan	4.6			

Defence spending
$bn, 2006

1	United States	535.9	16	Turkey	11.6
2	China	121.9	17	Israel	11.0
3	Russia	70.0	18	Netherlands	9.9
4	United Kingdom	55.4	19	United Arab Emirates	9.5
5	France	54.0	20	Indonesia[a]	8.4
6	Japan	41.1	21	Taiwan	7.7
7	Germany	37.8	22	Greece	7.3
8	Italy	30.6	23	Iran	7.2
9	Saudi Arabia	29.5	24	Myanmar	6.9
10	South Korea	24.6	25	Singapore	6.3
11	India	22.4	26	Poland	6.2
12	Australia	17.2	27	Ukraine	6.0
13	Brazil	16.2	28	Sweden	5.8
14	Canada	15.0	29	Colombia	5.4
15	Spain	14.4	30	Norway	5.0

Armed forces
'000, 2006

		Regulars	Reserves			Regulars	Reserves
1	China	2,105	800	12	Brazil	368	1,340
2	United States	1,498	1,083	13	Thailand	306	200
3	India	1,288	1,155	14	Indonesia	302	400
4	Russia	1,027	20,000	15	Syria	293	314
5	South Korea	687	4,500	16	Taiwan	290	1,657
6	Pakistan	619		17	France	255	25
7	Iran	545	350	18	Colombia	254	62
8	Turkey	510	379	19	Mexico	249	40
9	Egypt	469	479	20	Germany	246	162
10	Vietnam	455	5,000	21	Japan	240	42
11	Myanmar	406		22	Saudi Arabia	224	

a 2005

Environment

Environmental performance index[a], 2008

Highest			Lowest		
1	Switzerland	95.5	1	Niger	39.1
2	Norway	93.1	2	Angola	39.5
	Sweden	93.1	3	Sierra Leone	40.0
4	Finland	91.4	4	Mauritania	44.2
5	Costa Rica	90.5	5	Burkina Faso	44.3
6	Austria	89.4		Mali	44.3
7	New Zealand	88.9	7	Chad	45.9
8	Latvia	88.8	8	Congo-Kinshasa	47.3
9	Colombia	88.3	9	Guinea-Bissau	49.7
10	France	87.8		Yemen	49.7
11	Iceland	87.6	11	Guinea	51.3
12	Canada	86.6	12	Cambodia	53.8
13	Germany	86.3	13	Iraq	53.9
	Slovenia	86.3	14	Mozambique	53.9
15	United Kingdom	86.3	15	Madagascar	54.6
16	Lithuania	86.2	16	Burundi	54.7
17	Slovakia	86.0	17	Rwanda	54.9
18	Portugal	85.8	18	Zambia	55.1
19	Estonia	85.2	19	Sudan	55.5
20	Croatia	84.6	20	Central African Rep	56.0
21	Japan	84.5	21	Benin	56.1
22	Ecuador	84.4	22	Nigeria	56.2
23	Hungary	84.2	23	Bangladesh	58.0
	Italy	84.2	24	Pakistan	58.7
25	Albania	84.0	25	Ethiopia	58.8
	Denmark	84.0	26	Eritrea	59.4
	Malaysia	84.0	27	Malawi	59.9
28	Russia	83.9	28	India	60.3
29	Chile	83.4	29	Haiti	60.7
30	Luxembourg	83.1	30	Swaziland	61.3
	Panama	83.1	31	Uganda	61.6
	Spain	83.1	32	Togo	62.3
33	Dominican Republic	83.0	33	Senegal	62.8
34	Brazil	82.7	34	Cameroon	63.8
	Ireland	82.7	35	Tanzania	63.9
36	Uruguay	82.3	36	United Arab Emirates	64.0
37	Georgia	82.2	37	Kuwait	64.5
38	Argentina	81.8	38	Bolivia	64.7
39	United States	81.0	39	Papua New Guinea	64.8
40	Taiwan	80.8	40	Kazakhstan	65.0
41	Cuba	80.7		Uzbekistan	65.0
42	Belarus	80.5	42	China	65.1
	Poland	80.5		Myanmar	65.1
44	Greece	80.2	44	Côte d'Ivoire	65.2
45	Venezuela	80.0	45	Indonesia	66.2
46	Australia	79.8	46	Laos	66.3
	Mexico	79.8	47	Mongolia	68.1
48	Bosnia	79.7	48	Syria	68.2

a Based on a range of factors including environmental health, biodiversity, air pollution, water use, agricultural methods, tackling climate change.

Slum population[a]

% of total urban population, 2005

1	Afghanistan[b]	99	25	Uganda	67
2	Somalia[b]	97	26	Equatorial Guinea	66
3	Central African Rep	94		Malawi	66
	Ecuador	94		Nigeria	66
	Mauritania[b]	94		Tanzania	66
	Sudan	94	30	Yemen	65
7	Chad	91	31	Burundi	64
8	Angola	86	32	Belize[b]	62
	Haiti[b]	86		Togo	62
10	Myanmar	84	34	Botswana[b]	61
11	Guinea-Bissau	83		Jamaica	61
12	Ethiopia	82		Nepal	61
13	Madagascar	81		Oman[b]	61
14	Mozambique	80	38	West Bank and Gaza[b]	60
15	Laos	79	39	Mongolia	58
16	Burkina Faso[b]	77	40	Benin	57
	Sierra Leone	77		Iraq	57
18	Congo-Kinshasa	76		Zambia	57
19	Cambodia[b]	72	43	Côte d'Ivoire	56
	Pakistan	72		Liberia[b]	56
	Rwanda	72	45	Kenya	55
22	Bangladesh	71	46	Congo-Brazzaville	53
23	Cape Verde[b]	70	47	Bolivia	50
	Eritrea[b]	70		Lebanon[b]	50

Air pollution in cities

Micrograms of particulate matter per cubic metre, 2004

1	Cairo, Egypt	169	20	Nanchang, China	78
2	Delhi, India	150	21	Harbin, China	77
3	Kolkata, India	128	22	Changchun, China	74
4	Tianjin, China	125		Zibo, China	74
5	Chongqing, China	123	24	Shanghai, China	73
6	Kanpur, India	109	25	Guiyang, China	70
	Lucknow, India	109		Kunming, China	70
8	Jakarta, Indonesia	104	27	Qingdao, China	68
9	Shenyang, China	101	28	Pingxiang, China	67
10	Zhengzhou, China	97	29	Guanzhou, China	63
11	Jinan, China	94		Mumbai, India	63
12	Lanzhou, China	91	31	Sofia, Bulgaria	61
13	Beijing, China	89		Santiago, Chile	61
14	Taiyuan, China	88	33	Liupanshui, China	59
15	Chengdu, China	86	34	Córdoba, Argentina	58
16	Ahmadabad, India	83		Tehran, Iran	58
17	Anshan, China	82	36	Wulumqi, China	57
18	Wuhan, China	79	37	Nagpur, India	56
	Bangkok, Thailand	79	38	Istanbul, Turkey	55

a Urban households lacking one or more of: durable housing, secure tenancy, sufficient living area, access to improved water source, access to improved sanitation.
b 2001
c Suspended particulates that are less than 10 microns in diameter.

Biggest emitters of carbon dioxide
Millions of tonnes, 2004

1	United States	6,044.0	26	Malaysia	177.4
2	China	5,005.7	27	Venezuela	172.5
3	Russia	1,523.6	28	Egypt	158.1
4	India	1,341.8	29	United Arab Emirates	149.1
5	Japan	1,256.8	30	Netherlands	141.9
6	Germany	808.0	31	Argentina	141.7
7	Canada	638.8	32	Uzbekistan	137.8
8	United Kingdom	586.7	33	Pakistan	125.6
9	South Korea	465.2	34	Czech Republic	116.9
10	Italy	449.5	35	Nigeria	113.9
11	Mexico	437.6	36	Belgium	100.6
12	South Africa	436.6	37	Kuwait	99.3
13	Iran	433.2	38	Vietnam	98.6
14	Indonesia	377.9	39	Greece	96.6
15	France	373.4	40	Romania	90.3
16	Brazil	331.5	41	Norway	87.5
17	Spain	330.2	42	Iraq	81.6
18	Ukraine	329.7	43	Philippines	80.4
19	Australia	326.5	44	North Korea	79.0
20	Saudi Arabia	308.1	45	Israel	71.2
21	Poland	307.0	46	Austria	69.8
22	Thailand	267.8	47	Syria	68.4
23	Turkey	225.9	48	Finland	65.7
24	Kazakhstan	200.1	49	Belarus	64.8
25	Algeria	193.8	50	Chile	62.4

Largest amount of carbon dioxide emitted per person
Tonnes, 2004

1	Kuwait	40.4	23	Belgium	9.7
2	United Arab Emirates	37.8		South Korea	9.7
3	Trinidad & Tobago	24.7	25	South Africa	9.4
4	United States	20.6	26	Greece	8.7
5	Canada	20.0		Netherlands	8.7
6	Norway	19.1		Turkmenistan	8.7
7	Australia	16.2	29	Austria	8.5
8	Estonia	14.0	30	Slovenia	8.1
9	Saudi Arabia	13.7	31	Poland	8.0
10	Kazakhstan	13.3	32	Italy	7.7
11	Finland	12.6		New Zealand	7.7
12	Oman	12.5		Spain	7.7
13	Singapore	12.3	35	Malaysia	7.0
14	Czech Republic	11.5	36	Ukraine	6.9
15	Russia	10.6	37	Slovakia	6.7
16	Israel	10.5	38	Belarus	6.6
17	Ireland	10.4		Serbia	6.6
18	Libya	10.3		Venezuela	6.6
19	Denmark	9.8	41	Iran	6.4
	Germany	9.8	42	France	6.2
	Japan	9.8	43	Algeria	6.0
	United Kingdom	9.8	44	Sweden	5.9

Average annual % change in carbon emissions

Biggest increase, 1990–2004		*Biggest decrease, 1990–2004*	
1 Namibia	57.6	1 North Korea	-11.2
2 Laos	15.8	2 Tajikistan	-11.1
3 Bosnia	12.5	3 Afghanistan	-11.0
4 Kuwait	12.2	4 Georgia	-10.2
5 Vietnam	11.9	5 Moldova	-9.1
6 Swaziland	11.5	6 Gabon	-9.0
7 Nepal	10.3	7 Kyrgyzstan	-6.3
8 United Arab Emirates	9.3	8 Congo-Kinshasa	-6.1
9 Algeria	8.8	Latvia	-6.1
Congo-Brazzaville	8.8	10 Ukraine	-5.6
11 Madagascar	8.6	11 Lithuania	-4.7
Oman	8.6	12 Azerbaijan	-4.5
13 Sri Lanka	8.5	13 Puerto Rico	-4.1
14 Togo	8.3	14 Belarus	-3.8
15 Ethiopia	8.1	Kazakhstan	-3.8
16 Benin	7.9	16 Romania	-3.7
17 Honduras	7.7	17 Bulgaria	-3.4
18 Norway	7.4	18 Estonia	-3.3
19 Uganda	7.1	19 Serbia	-3.1
20 Haiti	7.0	Zimbabwe	-3.1
21 Bangladesh	6.9	21 Russia	-2.7
Malaysia	6.9	22 Mongolia	-2.1
23 Nigeria	6.8	23 Albania	-2.0
24 Guatemala	6.7	Czech Republic	-2.0
25 Mauritius	6.3	25 Cuba	-1.8
Thailand	6.3	Slovakia	-1.8
27 Myanmar	6.2	27 Zambia	-1.6
28 Dominican Republic	6.1	28 Poland	-1.2
29 Sudan	5.9	29 Mauritania	-1.1
30 El Salvador	5.8	Saudi Arabia	-1.1

Clean energy[a]

As % of total energy use, 2005

1 Sweden	48.2	Philippines	20.7
2 France	44.3	18 Canada	20.3
3 Kyrgyzstan	43.8	19 Uruguay	19.9
4 Tajikistan	41.5	20 Albania	19.2
5 Costa Rica	40.6	21 South Korea	18.0
6 Norway	36.4	22 Georgia	17.0
7 Armenia	33.6	23 Ukraine	16.9
8 Syria	32.9	24 Japan	16.8
9 Lithuania	32.4	25 Brazil	15.1
10 Slovakia	27.0	26 Czech Republic	14.8
11 Bulgaria	26.3	27 Hungary	13.4
12 Slovenia	25.0	28 Germany	12.9
13 New Zealand	23.4	29 Peru	12.4
14 El Salvador	22.6	30 Panama	12.3
15 Belgium	21.9	31 Colombia	12.0
16 Finland	20.7	32 Spain	11.5

a Energy that does not produce carbon dioxide when generated.

Urban population with access to improved sanitation
Lowest, %, 2004

1	Chad	24		Somalia	48
2	Ghana	27	17	Liberia	49
3	Congo-Brazzaville	28		Mauritania	49
4	Guinea	31	19	Namibia	50
5	Eritrea	32		Sudan	50
6	Gabon	37	21	Bangladesh	51
7	Burkina Faso	42	22	Cambodia	53
	Congo-Kinshasa	42		Mozambique	53
9	Niger	43		Nigeria	53
10	Ethiopia	44		Sierra Leone	53
11	Côte d'Ivoire	46		Tanzania	53
	Kenya	46	27	Uganda	54
13	Burundi	47	28	Angola	56
	Central African Rep	47		Nicaragua	56
15	Madagascar	48		Rwanda	56

Rural population with access to improved sanitation
Lowest, %, 2004

1	Eritrea	3	16	Togo	15
2	Chad	4	17	Angola	16
	Niger	4	18	Mozambique	19
4	Burkina Faso	6	19	Laos	20
5	Ethiopia	7	20	Bolivia	22
	Liberia	7		India	22
7	Cambodia	8	22	Guinea-Bissau	23
	Mauritania	8	23	Sudan	24
9	Benin	11	24	Botswana	25
	Ghana	11		Congo-Brazzaville	25
	Guinea	11		Congo-Kinshasa	25
12	Central African Rep	12	27	Madagascar	26
13	Namibia	13	28	China	28
14	Haiti	14		Yemen	28
	Somalia	14	30	Côte d'Ivoire	29

Rural population with access to improved water source
Lowest, %, 2004

1	Ethiopia	11		Togo	36
2	Romania	16	16	Angola	40
3	Mozambique	26		Zambia	40
4	Congo-Brazzaville	27	18	Chad	43
	Somalia	27		Laos	43
6	Congo-Kinshasa	29	20	Cameroon	44
7	Mongolia	30		Mauritania	44
8	Nigeria	31	22	Kenya	46
9	Papua New Guinea	32		Sierra Leone	46
10	Cambodia	35	24	Gabon	47
	Guinea	35	25	Tajikistan	48
	Madagascar	35	26	Guinea-Bissau	49
13	Mali	36		Tanzania	49
	Niger	36	28	Liberia	52

Country profiles

ALGERIA

Area	2,381,741 sq km	Capital	Algiers
Arable as % of total land	3	Currency	Algerian dinar (AD)

People

Population	33.4m	Life expectancy: men		70.9 yrs
Pop. per sq km	14.0		women	73.7 yrs
Av. ann. growth		Adult literacy		75.4%
in pop. 2010–15	1.45%	Fertility rate (per woman)		2.3
Pop. under 15	29.6%	Urban population		60.0%
Pop. over 60	12.3%			per 1,000 pop.
No. of men per 100 women	102	Crude birth rate		21
Human Development Index	73.3	Crude death rate		4.9

The economy

GDP	AD8,382bn	GDP per head	$3,430
GDP	$115bn	GDP per head in purchasing	
Av. ann. growth in real		power parity (USA=100)	14.4
GDP 1997–2007	4.6%	Economic freedom index	55.7

Origins of GDP		Components of GDP	
	% of total		% of total
Agriculture	8	Private consumption	31.6
Industry, of which:	61	Public consumption	11.3
manufacturing	6	Investment	22.8
Services	30	Exports	49.3
		Imports	-22.1

Structure of employment

	% of total		% of labour force
Agriculture	21	Unemployed 2006	12.3
Industry	24	Av. ann. rate 1995–2006	24.6
Services	55		

Energy

	m TOE		
Total output	175.1	Net energy imports as %	
Total consumption	34.8	of energy use	-404
Consumption per head,			
kg oil equivalent	1,058		

Inflation and finance

Consumer price		av. ann. increase 2001–06	
inflation 2007	3.5%	Narrow money (M1)	20.3%
Av. ann. inflation 2002–07	2.8%	Broad money	14.8%
Money market rate, 2007	3.13%		

Exchange rates

	end 2007		December 2007
AD per $	66.83	Effective rates	2000 = 100
AD per SDR	105.61	– nominal	83.39
AD per €	98.38	– real	80.83

Trade

Principal exports		Principal imports	
	$bn fob		*$bn cif*
Hydrocarbons	53.6	Capital goods	8.0
Semi-finished goods	0.8	Semi-finished goods	4.6
Raw materials	0.2	Food	3.4
Total incl. others	**54.7**	**Total incl. others**	**22.3**

Main export destinations		Main origins of imports	
	% of total		*% of total*
United States	26.7	France	24.8
Italy	16.7	Italy	9.7
Spain	9.2	China	9.6
France	8.6	Germany	6.6

Balance of payments, reserves and debt, $bn

Visible exports fob	54.7	Change in reserves	22.3
Visible imports fob	-20.7	Level of reserves	
Trade balance	34.1	end Dec.	81.5
Invisibles inflows	5.0	No. months of import cover	30.2
Invisibles outflows	-11.5	Official gold holdings, m oz	5.6
Net transfers	1.6	Foreign debt	6
Current account balance	29.0	– as % of GDP	5
– as % of GDP	25.2	– as % of total exports	10
Capital balance	-11.2	Debt service ratio[a]	12
Overall balance	17.7		

Health and education

Health spending, % of GDP	3.5	Education spending, % of GDP	...
Doctors per 1,000 pop.	1.3	Enrolment, %: primary	110
Hospital beds per 1,000 pop.	1.7	secondary	83
Improved-water source access,		tertiary	22
% of pop.	85		

Society

No. of households	5.6m	Colour TVs per 100 households	74.0
Av. no. per household	5.9	Telephone lines per 100 pop.	8.5
Marriages per 1,000 pop.	8.7	Mobile telephone subscribers	
Divorces per 1,000 pop.	...	per 100 pop.	63.0
Cost of living, Dec. 2007		Computers per 100 pop.	1.1
New York = 100	57	Internet hosts per 1,000 pop.	...

a 2005

ARGENTINA

Area	2,766,889 sq km	Capital	Buenos Aires
Arable as % of total land	10	Currency	Peso (P)

People

Population	39.1m	Life expectancy: men	71.6 yrs
Pop. per sq km	14.1	women	79.1 yrs
Av. ann. growth		Adult literacy	97.2%
in pop. 2010–15	0.93%	Fertility rate (per woman)	2.2
Pop. under 15	26.4%	Urban population	90.6%
Pop. over 60	13.9%		per 1,000 pop.
No. of men per 100 women	96	Crude birth rate	19
Human Development Index	86.9	Crude death rate	7.7

The economy

GDP	P655bn	GDP per head	$5,480
GDP	$214bn	GDP per head in purchasing	
Av. ann. growth in real		power parity (USA=100)	27.3
GDP 1997–2007	2.9%	Economic freedom index	55.1

Origins of GDP		Components of GDP	
	% of total		% of total
Agriculture	8.5	Private consumption	59.0
Industry, of which:	35.9	Public consumption	12.4
manufacturing	22.4	Investment	23.4
Services	55.6	Exports	24.8
		Imports	-19.2

Structure of employment

	% of total		% of labour force
Agricultural	1	Unemployed 2006	9.5
Industry	24	Av. ann. rate 1995–2006	15.0
Services	75		

Energy

	m TOE		
Total output	81	Net energy imports as %	
Total consumption	63.7	of energy use	-27
Consumption per head			
kg oil equivalent	1,644		

Inflation and finance

		av. ann. increase 2001–06	
Consumer price			
inflation 2007	8.8%	Narrow money (M1)	37.7%
Av. ann. inflation 2002–07	9.4%	Broad money	22.5%
Money market rate, 2007	8.67%		

Exchange rates

	end 2007		December 2007
P per $	3.13	Effective rates	2000 = 100
P per SDR	4.94	– nominal	...
P per €	4.61	– real	...

Trade

Principal exports		Principal imports	
	$bn fob		*$bn cif*
Agricultural products	15.2	Intermediate goods	11.9
Manufactures	14.8	Capital goods	8.2
Primary products	8.6	Consumer goods	4.0
Fuels	7.8	Fuels	1.7
Total incl. others	**46.5**	Total incl. others	**34.2**

Main export destinations		Main origins of imports	
	% of total		*% of total*
Brazil	16.8	Brazil	37.2
Chile	8.8	United States	15.4
United States	8.3	China	6.5
China	7.2	Germany	5.2

Balance of payments, reserves and debt, $bn

Visible exports fob	46.5	Change in reserves	3.9
Visible imports fob	-32.6	Level of reserves	
Trade balance	13.9	end Dec.	32.0
Invisibles inflows	13.1	No. months of import cover	7.4
Invisibles outflows	-19.4	Official gold holdings, m oz	1.8
Net transfers	0.5	Foreign debt	122.2
Current account balance	8.1	– as % of GDP	6.8
– as % of GDP	3.8	– as % of total exports	230
Capital balance	4.3	Debt service ratio	32
Overall balance	13.3		

Health and education

Health spending, % of GDP	10.2	Education spending, % of GDP	3.8
Doctors per 1,000 pop.	3.0	Enrolment, %: primary	112
Hospital beds per 1,000 pop.	...	secondary	84
Improved-water source access,		tertiary	64
% of pop.	96		

Society

No. of households	10.5m	Colour TVs per 100 households	94.2
Av. no. per household	3.6	Telephone lines per 100 pop.	24.2
Marriages per 1,000 pop.	3.2	Mobile telephone subscribers	
Divorces per 1,000 pop.	...	per 100 pop.	80.5
Cost of living, Dec. 2007		Computers per 100 pop.	9.0
New York = 100	57	Internet hosts per 1,000 pop.	80.0

AUSTRALIA

Area	7,682,300 sq km	Capital	Canberra
Arable as % of total land	6	Currency	Australian dollar (A$)

People

Population	20.4m	Life expectancy: men	78.9 yrs
Pop. per sq km	2.7	women	83.6 yrs
Av. ann. growth		Adult literacy	...
in pop. 2010–15	0.95%	Fertility rate (per woman)	1.8
Pop. under 15	19.5%	Urban population	92.7%
Pop. over 60	17.8%		per 1,000 pop.
No. of men per 100 women	99	Crude birth rate	13
Human Development Index	96.2	Crude death rate	7.1

The economy

GDP	A$1,036bn	GDP per head	$38,260
GDP	$781bn	GDP per head in purchasing	
Av. ann. growth in real		power parity (USA=100)	80.8
GDP 1997–2007	3.9%	Economic freedom index	82.0

Origins of GDP		Components of GDP	
	% of total		% of total
Agriculture	2.8	Private consumption	56.3
Industry, of which:	28.1	Public consumption	18.3
manufacturing	11.0	Investment	26.7
Services	69.1	Exports	20.9
		Imports	-22.1

Structure of employment

	% of total		% of labour force
Agriculture	4	Unemployed 2006	5.0
Industry	21	Av. ann. rate 1995–2006	6.8
Services	75		

Energy

	m TOE		
Total output	271	Net energy imports as %	
Total consumption	122	of energy use	-122
Consumption per head,			
kg oil equivalent	5,978		

Inflation and finance

Consumer price		av. ann. increase 2001–06	
inflation 2007	2.3%	Narrow money (M1)	12.3%
Av. ann. inflation 2002–07	2.7%	Broad money	10.6%
Money market rate, 2007	6.39%	Household saving rate, 2007	0.3%

Exchange rates

	end 2007		December 2007
A$ per $	1.13	Effective rates	2000 = 100
A$ per SDR	1.79	– nominal	125.60
A$ per €	1.66	– real	134.60

Trade

Principal exports		Principal imports	
	$bn fob		*$bn cif*
Coal	17.5	Intermediate & other goods	61.5
Meat & meat products	5.3	Consumption goods	40.3
Wheat	3.3	Capital goods	40.0
Total incl. others	**123.5**	Total incl. others	**132.8**

Main export destinations		Main origins of imports	
	% of total		*% of total*
Japan	19.3	China	15.9
China	12.1	United States	15.6
South Korea	7.4	Japan	10.6
United States	6.1	Singapore	6.6
EU25	12.4	EU25	20.7

Balance of payments, reserves and aid, $bn

Visible exports fob	124.9	Overall balance	9.7
Visible imports fob	-134.5	Change in reserves	11.8
Trade balance	-9.6	Level of reserves	
Invisibles inflows	55.0	end Dec.	55.1
Invisibles outflows	-86.3	No. months of import cover	3.0
Net transfers	-0.2	Official gold holdings, m oz	2.6
Current account balance	-41.0	Aid given	2.12
– as % of GDP	-5.3	– as % of GDP	0.27
Capital balance	51.2		

Health and education

Health spending, % of GDP	8.8	Education spending, % of GDP	4.5
Doctors per 1,000 pop.	2.5	Enrolment, %: primary	105
Hospital beds per 1,000 pop.	4.0	secondary	150
Improved-water source access,		tertiary	73
% of pop.	100		

Society

No. of households	7.7m	Colour TVs per 100 households	98.8
Av. no. per household	2.7	Telephone lines per 100 pop.	48.8
Marriages per 1,000 pop.	5.5	Mobile telephone subscribers	
Divorces per 1,000 pop.	2.5	per 100 pop.	97.0
Cost of living, Dec. 2007		Computers per 100 pop.	75.7
New York = 100	116	Internet hosts per 1,000 pop.	524.9

AUSTRIA

Area	83,855 sq km	Capital	Vienna
Arable as % of total land	17	Currency	Euro (€)

People

Population	8.2m	Life expectancy: men	76.9 yrs
Pop. per sq km	97.8	women	82.6 yrs
Av. ann. growth		Adult literacy	...
in pop. 2010–15	0.17%	Fertility rate (per woman)	1.5
Pop. under 15	15.8%	Urban population	65.8%
Pop. over 60	21.9%		per 1,000 pop.
No. of men per 100 women	96	Crude birth rate	9
Human Development Index	94.8	Crude death rate	9.4

The economy

GDP	€257bn	GDP per head	$37,270
GDP	$322bn	GDP per head in purchasing	
Av. ann. growth in real		power parity (USA=100)	82.0
GDP 1997–2007	2.7%	Economic freedom index	70.0

Origins of GDP		**Components of GDP**	
	% of total		% of total
Agriculture	1.7	Private consumption	55.3
Industry, of which:	30.6	Public consumption	18.0
manufacturing	...	Investment	20.6
Services	67.7	Exports	56.3
		Imports	-50.3

Structure of employment

	% of total		% of labour force
Agriculture	6	Unemployed 2006	4.7
Industry	28	Av. ann. rate 1995–2006	4.2
Services	66		

Energy

	m TOE		
Total output	9.8	Net energy imports as %	
Total consumption	34.4	of energy use	71
Consumption per head,			
kg oil equivalent	4,174		

Inflation and finance

		av. ann. increase 2001–06	
Consumer price			
inflation 2007	2.2%	Euro area:	
Av. ann. inflation 2002–07	1.9%	Narrow money (M1)	10.5%
Deposit rate, h'holds, 2007	3.16%	Broad money	7.4%
		Household saving rate, 2007	10.6%

Exchange rates

	end 2007		December 2007
€ per $	0.68	Effective rates	2000 = 100
€ per SDR	1.07	– nominal	108.10
		– real	102.70

Trade

Principal exports		Principal imports	
	$bn fob		*$bn cif*
Consumer goods	59.8	Consumer goods	54.7
Investment goods	35.3	Investment goods	28.7
Intermediate goods	19.8	Raw materials (incl. fuels)	24.5
Raw materials (incl. fuels)	11.2	Intermediate goods	18.9
Food & beverages	7.8	Food & beverages	7.3
Total incl. others	**130.4**	Total incl. others	**131.0**

Main export destinations		Main origins of imports	
	% of total		*% of total*
Germany	31.8	Germany	42.5
Eastern Europe	19.6	Eastern Europe	14.4
Italy	8.7	Italy	6.7
United States	5.8	Switzerland	3.7
EU25	70.5	EU25	74.2

Balance of payments, reserves and aid, $bn

Visible exports fob	134.3	Overall balance	-0.8
Visible imports fob	-133.7	Change in reserves	1.1
Trade balance	0.6	Level of reserves	
Invisibles inflows	72.7	end Dec.	12.9
Invisibles outflows	-61.7	No. months of import cover	0.8
Net transfers	-1.3	Official gold holdings, m oz	9.3
Current account balance	10.3	Aid given	1.50
– as % of GDP	3.2	– as % of GDP	0.47
Capital balance	-10.0		

Health and education

Health spending, % of GDP	10.2	Education spending, % of GDP	5.4
Doctors per 1,000 pop.	3.7	Enrolment, %: primary	102
Hospital beds per 1,000 pop.	7.7	secondary	102
Improved-water source access,		tertiary	50
% of pop.	100		

Society

No. of households	3.5m	Colour TVs per 100 households	98.7
Av. no. per household	2.4	Telephone lines per 100 pop.	43.4
Marriages per 1,000 pop.	4.9	Mobile telephone subscribers	
Divorces per 1,000 pop.	2.3	per 100 pop.	112.8
Cost of living, Dec. 2007		Computers per 100 pop.	60.7
New York = 100	119	Internet hosts per 1,000 pop.	315.8

BANGLADESH

Area	143,998 sq km	Capital	Dhaka
Arable as % of total land	61	Currency	Taka (Tk)

People

Population	144.4m	Life expectancy: men	63.2 yrs
Pop. per sq km	1,002.8	women	65.0 yrs
Av. ann. growth		Adult literacy	53.5%
in pop. 2010–15	1.56%	Fertility rate (per woman)	2.6
Pop. under 15	35.2%	Urban population	25.0%
Pop. over 60	5.7%		per 1,000 pop.
No. of men per 100 women	105	Crude birth rate	27
Human Development Index	54.7	Crude death rate	7.5

The economy

GDP	Tk4,157bn	GDP per head	$430
GDP	$61.9bn	GDP per head in purchasing	
Av. ann. growth in real		power parity (USA=100)	2.6
GDP 1996–2006	6.2%	Economic freedom index	44.9

Origins of GDPa		Components of GDPa	
	% of total		% of total
Agriculture	19.7	Private consumption	74.1
Industry, of which:	28.0	Public consumption	5.8
manufacturing	17.3	Investment	25.6
Services	52.3	Exports	16.4
		Imports	-24.2

Structure of employment

	% of total		% of labour force
Agriculture	52	Unemployed 2003	4.3
Industry	14	Av. ann. rate 1995–2003	2.7
Services	35		

Energy

			m TOE
Total output	19.3	Net energy imports as %	
Total consumption	24.2	of energy use	20
Consumption per head,			
kg oil equivalent	158		

Inflation and finance

Consumer price		av. ann. increase 2001–06	
inflation 2007	9.1%	Narrow money (M1)	15.7%
Av. ann. inflation 2002–07	7.6%	Broad money	15.9%
Deposit rate, 2007	9.18%		

Exchange rates

	end 2007		December 2007
Tk per $	68.58	Effective rates	2000 = 100
Tk per SDR	108.37	– nominal	...
Tk per €	100.96	– real	...

Trade

Principal exports[a]		Principal imports[a]	
	$bn fob		*$bn cif*
Clothing	6.0	Textiles & yarn	3.4
Fish & fish products	0.4	Capital goods	3.0
Jute goods	0.3	Fuels	2.0
Leather	0.3	Iron & steel	0.9
Total incl. others	**9.3**	Total incl. others	**14.8**

Main export destinations		Main origins of imports	
	% of total		*% of total*
United States	25.2	China	18.0
Germany	12.7	India	12.7
United Kingdom	9.9	Kuwait	8.0
France	5.5	Singapore	5.6
Italy	3.9	Japan	3.8

Balance of payments, reserves and debt, $bn

Visible exports fob	11.6	Change in reserves	1.1
Visible imports fob	-14.4	Level of reserves	
Trade balance	-2.9	end Dec.	3.9
Invisibles inflows	1.5	No. months of import cover	2.6
Invisibles outflows	-3.4	Official gold holdings, m oz	0.1
Net transfers	5.9	Foreign debt	20.5
Current account balance	1.2	– as % of GDP	22
– as % of GDP	1.9	– as % of total exports	91
Capital balance	0.3	Debt service ratio	4
Overall balance	0.9		

Health and education

Health spending, % of GDP	2.8	Education spending, % of GDP	2.7
Doctors per 1,000 pop.	0.3	Enrolment, %: primary	103
Hospital beds per 1,000 pop.	3.0	secondary	44
Improved-water source access,		tertiary	6
% of pop.	74		

Society

No. of households	25.7m	Colour TVs per 100 households	2.7
Av. no. per household	6.0	Telephone lines per 100 pop.	0.8
Marriages per 1,000 pop.	...	Mobile telephone subscribers	
Divorces per 1,000 pop.	...	per 100 pop.	13.3
Cost of living, Dec. 2007		Computers per 100 pop.	2.2
New York = 100	53	Internet hosts per 1,000 pop.	...

a Fiscal year ending June 30 2006.

BELGIUM

Area	30,520 sq km	Capital	Brussels
Arable as % of total land	28	Currency	Euro (€)

People

Population	10.4m	Life expectancy: men	76.5 yrs
Pop. per sq km	340.8	women	82.3 yrs
Av. ann. growth		Adult literacy	...
in pop. 2010–15	0.17%	Fertility rate (per woman)	1.7
Pop. under 15	17.0%	Urban population	97.3%
Pop. over 60	22.1%		per 1,000 pop.
No. of men per 100 women	96	Crude birth rate	11
Human Development Index	94.6	Crude death rate	10.0

The economy

GDP	€314bn	GDP per head	$37,890
GDP	$394bn	GDP per head in purchasing	
Av. ann. growth in real		power parity (USA=100)	76.3
GDP 1997–2007	2.5%	Economic freedom index	71.5

Origins of GDP		Components of GDP	
	% of total		% of total
Agriculture	0.9	Private consumption	52.5
Industry, of which:	24.2	Public consumption	22.3
manufacturing	...	Investment	20.8
Services	74.9	Exports	87.5
		Imports	-84.5

Structure of employment

	% of total		% of labour force
Agriculture	2	Unemployed 2006	8.2
Industry	25	Av. ann. rate 1995–2006	8.2
Services	73		

Energy

	m TOE		
Total output	13.8	Net energy imports as %	
Total consumption	56.7	of energy use	75
Consumption per head,			
kg oil equivalent	5,407		

Inflation and finance

		av. ann. increase 2001–06	
Consumer price inflation 2007	1.8%	Euro area:	
Av. ann. inflation 2002–07	2.0%	Narrow money (M1)	10.5%
Treasury bill rate, 2007	3.80%	Broad money	7.4%
		Household saving rate, 2007	10.0%

Exchange rates

	end 2007		December 2007
€ per $	0.68	Effective rates	2000 = 100
€ per SDR	1.07	– nominal	113.10
		– real	120.40

Trade

Principal exports		Principal imports	
	$bn fob		*$bn cif*
Chemicals & related products	106.8	Chemicals & related products	88.0
Machinery & transport equip.	90.5	Machinery & transport equip.	87.9
Minerals, fuels & lubricants	29.5	Minerals, fuels & lubricants	48.9
Food, drink & tobacco	28.7	Agriculture, food & drink	24.2
Total incl. others	**367.0**	Total incl. others	**352.0**

Main export destinations		Main origins of imports	
	% of total		*% of total*
Germany	20.0	Netherlands	18.4
France	17.0	Germany	17.6
Netherlands	12.1	France	11.3
United Kingdom	8.0	United Kingdom	6.7
EU25	76.9	EU25	75.0

Balance of payments, reserves and aid, $bn

Visible exports fob	281.1	Overall balance	0.2
Visible imports fob	-277.8	Change in reserves	1.4
Trade balance	3.4	Level of reserves	
Invisibles inflows	129.9	end Dec.	13.4
Invisibles outflows	-115.9	No. months of import cover	0.4
Net transfers	-6.7	Official gold holdings, m oz	7.3
Current account balance	10.7	Aid given	1.98
– as % of GDP	2.7	– as % of GDP	0.50
Capital balance	-12.3		

Health and education

Health spending, % of GDP	9.6	Education spending, % of GDP	6.0
Doctors per 1,000 pop.	4.2	Enrolment, %: primary	102
Hospital beds per 1,000 pop.	5.3	secondary	110
Improved-water source access, % of pop.	...	tertiary	63

Society

No. of households	4.5m	Colour TVs per 100 households	98.8
Av. no. per household	2.4	Telephone lines per 100 pop.	45.2
Marriages per 1,000 pop.	4.2	Mobile telephone subscribers	
Divorces per 1,000 pop.	3.0	per 100 pop.	92.6
Cost of living, Dec. 2007		Computers per 100 pop.	37.7
New York = 100	114	Internet hosts per 1,000 pop.	347.9

BRAZIL

Area	8,511,965 sq km	Capital	Brasilia
Arable as % of total land	7	Currency	Real (R)

People

Population	188.9m	Life expectancy: men	68.8 yrs
Pop. per sq km	22.2	women	76.1 yrs
Av. ann. growth		Adult literacy	88.6%
in pop. 2010–15	1.08%	Fertility rate (per woman)	2.2
Pop. under 15	27.8%	Urban population	84.2%
Pop. over 60	8.8%		per 1,000 pop.
No. of men per 100 women	97	Crude birth rate	21
Human Development Index	80.0	Crude death rate	6.3

The economy

GDP	R2,323bn	GDP per head	$5,650
GDP	$1,068bn	GDP per head in purchasing	
Av. ann. growth in real		power parity (USA=100)	20.4
GDP 1997–2007	3.1%	Economic freedom index	55.9

Origins of GDP		Components of GDP	
	% of total		% of total
Agriculture	5.2	Private consumption	60.4
Industry, of which:	30.1	Public consumption	19.8
manufacturing	...	Investment	16.5
Services	64.7	Exports	14.6
		Imports	-11.7

Structure of employment

	% of total		% of labour force
Agriculture	21	Unemployed 2004	8.9
Industry	21	Av. ann. rate 1995–2004	8.5
Services	58		

Energy

			m TOE
Total output	187.8	Net energy imports as %	
Total consumption	209.5	of energy use	10
Consumption per head,			
kg oil equivalent	1,122		

Inflation and finance

Consumer price		av. ann. increase 2001–06	
inflation 2007	3.6%	Narrow money (M1)	15.8%
Av. ann. inflation 2002–07	7.1%	Broad money	17.2%
Money market rate, 2007	11.98%		

Exchange rates

	end 2007		December 2007
R per $	1.78	Effective rates	2000 = 100
R per sdr	2.80	– nominal	...
R per €	2.62	– real	...

Trade

Principal exports		Principal imports	
	$bn fob		*$bn cif*
Transport equipment & parts	20.1	Machines & electrical	
Metal goods	15.1	equipment	23.6
Soyabeans etc.	10.5	Oil & derivatives	15.2
Chemical products	3.9	Chemical products	14.4
		Transport equipment & parts	10.3
Total incl. others	**137.8**	Total incl. others	**95.9**

Main export destinations		Main origins of imports	
	% of total		*% of total*
United States	17.9	United States	16.3
Argentina	8.5	Argentina	8.8
China	6.1	China	8.7
Germany	4.1	Netherlands	0.9

Balance of payments, reserves and debt, $bn

Visible exports fob	137.8	Change in reserves	32.0
Visible imports fob	-91.4	Level of reserves	
Trade balance	46.5	end Dec.	85.8
Invisibles inflows	25.9	No. months of import cover	6.7
Invisibles outflows	-63.0	Official gold holdings, m oz	1.1
Net transfers	4.3	Foreign debt	194.2
Current account balance	13.6	– as % of GDP	26
– as % of GDP	1.3	– as % of total exports	158
Capital balance	16.0	Debt service ratio	37
Overall balance	30.6		

Health and education

Health spending, % of GDP	7.9	Education spending, % of GDP	4.0
Doctors per 1,000 pop.	1.0	Enrolment, %: primary	137
Hospital beds per 1,000 pop.	2.6	secondary	105
Improved-water source access,		tertiary	25
% of pop.	90		

Society

No. of households	50.7m	Colour TVs per 100 households	91.6
Av. no. per household	3.7	Telephone lines per 100 pop.	20.5
Marriages per 1,000 pop.	4.1	Mobile telephone subscribers	
Divorces per 1,000 pop.	0.8	per 100 pop.	52.9
Cost of living, Dec. 2007		Computers per 100 pop.	16.1
New York = 100	82	Internet hosts per 1,000 pop.	53.7

BULGARIA

Area	110,994 sq km	Capital	Sofia
Arable as % of total land	29	Currency	Lev (BGL)

People

Population	7.7m	Life expectancy: men		69.5 yrs
Pop. per sq km	69.4	women		76.7 yrs
Av. ann. growth		Adult literacy		98.2%
in pop. 2010–15	-0.80%	Fertility rate (per woman)		1.3
Pop. under 15	13.8%	Urban population		70.5%
Pop. over 60	29.9%			per 1,000 pop.
No. of men per 100 women	94	Crude birth rate		10
Human Development Index	82.4	Crude death rate		14.8

The economy

GDP	BGL49.1bn	GDP per head	$4,090
GDP	$31.5bn	GDP per head in purchasing	
Av. ann. growth in real		power parity (USA=100)	23.4
GDP 1997–2007	5.6%	Economic freedom index	62.9

Origins of GDP		Components of GDP	
	% of total		% of total
Agriculture	8.5	Private consumption	78.1
Industry, of which:	30.9	Public consumption	8.9
manufacturing	...	Investment	25.9
Services	59.5	Exports	64.5
		Imports	-83.3

Structure of employment

	% of total		% of labour force
Agriculture	9	Unemployed 2006	9.0
Industry	34	Av. ann. rate 1995–2006	14.3
Services	57		

Energy

		m TOE	
Total output	10.6	Net energy imports as %	
Total consumption	20.1	of energy use	47
Consumption per head,			
kg oil equivalent	2,592		

Inflation and finance

Consumer price		av. ann change 2001–06	
inflation 2007	8.4%	Narrow money (M1)	21.7%
Av. ann. inflation 2002–07	5.8%	Broad money	21.6%
Money market rate, 2007	4.03%		

Exchange rates

	end 2007		December 2007
BGL per $	1.33	Effective rates	2000 = 100
BGL per SDR	2.10	– nominal	114.76
BGL per €	1.96	– real	140.97

Trade

Principal exports		Principal imports	
	$bn fob		*$bn cif*
Other metals	2.2	Mineral fuels	4.1
Clothing	2.1	Machinery & equipment	2.1
Iron & steel	1.1	Textiles	1.8
Chemicals	0.8	Chemicals	1.5
Total incl. others	**15.1**	Total incl. others	**23.3**

Main export destinations		Main origins of imports	
	% of total		*% of total*
Italy	10.8	Germany	17.4
Turkey	10.1	Russia	12.5
Germany	9.8	Italy	8.8
Greece	8.1	Turkey	6.1

Balance of payments, reserves and debt, $bn

Visible exports fob	15.1	Change in reserves	3.1
Visible imports fob	-21.9	Level of reserves	
Trade balance	-6.8	end Dec.	11.8
Invisibles inflows	6.6	No. months of import cover	5.1
Invisibles outflows	-5.7	Official gold holdings, m oz	1.3
Net transfers	0.8	Foreign debt	20.9
Current account balance	-5.0	– as % of GDP	74
– as % of GDP	-15.9	– as % of total exports	110
Capital balance	7.0	Debt service ratio	12
Overall balance	2.3		

Health and education

Health spending, % of GDP	7.7	Education spending, % of GDP	4.5
Doctors per 1,000 pop.	3.7	Enrolment, %: primary	100
Hospital beds per 1,000 pop.	6.4	secondary	106
Improved-water source access,		tertiary	46
% of pop.	99		

Society

No. of households	2.9m	Colour TVs per 100 households	90.2
Av. no. per household	2.6	Telephone lines per 100 pop.	31.3
Marriages per 1,000 pop.	4.0	Mobile telephone subscribers	
Divorces per 1,000 pop.	2.2	per 100 pop.	107.6
Cost of living, Dec. 2007		Computers per 100 pop.	6.3
New York = 100	66	Internet hosts per 1,000 pop.	61.1

CAMEROON

Area	475,442 sq km	Capital	Yaoundé
Arable as % of total land	13	Currency	CFA franc (CFAfr)

People

Population	16.6m	Life expectancy: men		50.0 yrs
Pop. per sq km	34.9		women	50.8 yrs
Av. ann. growth		Adult literacy		67.9%
in pop. 2010–15	1.81%	Fertility rate (per woman)		3.8
Pop. under 15	41.8%	Urban population		52.9%
Pop. over 60	5.4%			per 1,000 pop.
No. of men per 100 women	100	Crude birth rate		37
Human Development Index	53.2	Crude death rate		14.4

The economy

GDP	CFAfr9,581bn	GDP per head	$1,100
GDP	$18.3bn	GDP per head in purchasing	
Av. ann. growth in real		power parity (USA=100)	4.8
GDP 1996–2006	4.3%	Economic freedom index	54.0

Origins of GDP		Components of GDP	
	% of total		% of total
Agriculture	20	Private consumption	72
Industry, of which:	33	Public consumption	11
manufacturing	18	Investment	18
Services	47	Exports	26
		Imports	-27

Structure of employment

	% of total		% of labour force
Agriculture	70	Unemployed 2006	...
Industry	13	Av. ann. rate 1995–2006	...
Services	17		

Energy

	m TOE		
Total output	11.9	Net energy imports as %	
Total consumption	7	of energy use	-71
Consumption per head,			
kg oil equivalent	392		

Inflation and finance

		av. ann. change 2001–06	
Consumer price			
inflation 2007	0.9%	Narrow money (M1)	5.6%
Av. ann. inflation 2002–07	1.7%	Broad money	7.6%
Deposit rate, 2007	4.25%		

Exchange rates

	end 2007		December 2007
CFAfr per $	445.59	Effective rates	2000 = 100
CFAfr per SDR	704.15	– nominal	114.50
CFAfr per €	655.95	– real	113.10

Trade

Principal exports[a]		Principal imports[a]	
	$bn fob		*$bn cif*
Crude oil	1.4	Capital goods	0.4
Cocoa	0.3	Intermediate goods	0.4
Cotton	0.1	Minerals & raw materials	0.4
Total incl. others	**2.9**	Total incl. others	**2.5**

Main export destinations		Main origins of imports	
	% of total		*% of total*
Spain	20.9	France	23.5
Italy	15.2	Nigeria	13.2
France	11.4	China	7.2
South Korea	7.6	United States	4.5

Balance of payments, reserves and debt, $bn

Visible exports fob	3.5	Change in reserves	0.8
Visible imports fob	-3.1	Level of reserves	
Trade balance	0.4	end Dec.	1.7
Net invisibles	-0.8	No. months of import cover	4.0
Net transfers	0.3	Official gold holdings, m oz	0.0
Current account balance	-0.1	Foreign debt	3.2
– as % of GDP	-0.6	– as % of GDP	18
Capital balance	0.7	– as % of total exports	70
Overall balance	0.7	Debt service ratio[a]	1

Health and education

Health spending, % of GDP	5.2	Education spending, % of GDP	3.3
Doctors per 1,000 pop.	0.2	Enrolment, %: primary	107
Hospital beds per 1,000 pop.	...	secondary	24
Improved-water source access,		tertiary	7
% of pop.	66		

Society

No. of households	4.5m	Colour TVs per 100 households	...
Av. no. per household	3.7	Telephone lines per 100 pop.	0.8
Marriages per 1,000 pop.	...	Mobile telephone subscribers	
Divorces per 1,000 pop.	...	per 100 pop.	18.9
Cost of living, Dec. 2007		Computers per 100 pop.	1.1
New York = 100	...	Internet hosts per 1,000 pop.	...

a 2005

CANADA

Area[a]	9,970,610 sq km	Capital	Ottawa
Arable as % of total land	5	Currency	Canadian dollar (C$)

People

Population	32.6m	Life expectancy: men	78.3 yrs
Pop. per sq km	3.3	women	82.9 yrs
Av. ann. growth		Adult literacy	...
in pop. 2010–15	0.84%	Fertility rate (per woman)	1.5
Pop. under 15	17.6%	Urban population	81.1%
Pop. over 60	17.8%		per 1,000 pop.
No. of men per 100 women	98	Crude birth rate	11
Human Development Index	96.1	Crude death rate	7.4

The economy

GDP	C$1,442bn	GDP per head	$39,010
GDP	$1,272bn	GDP per head in purchasing	
Av. ann. growth in real		power parity (USA=100)	83.5
GDP 1997–2007	3.7%	Economic freedom index	80.2

Origins of GDP

	% of total
Agriculture	2.3
Industry, of which:	29.1
manufacturing & mining	20.4
Services	68.6

Components of GDP

	% of total
Private consumption	55.6
Public consumption	19.3
Investment	22.0
Exports	36.3
Imports	-33.7

Structure of employment

	% of total		% of labour force
Agriculture	3	Unemployed 2006	6.3
Industry	22	Av. ann. rate 1995–2006	7.8
Services	75		

Energy

	m TOE		
Total output	401.3	Net energy imports as %	
Total consumption	272	of energy use	-48
Consumption per head,			
kg oil equivalent	8,417		

Inflation and finance

			av. ann. increase 2001–06
Consumer price			
inflation 2007	2.1%	Narrow money (M1)	6.8%
Av. ann. inflation 2002–07	2.2%	Broad money	6.1%
Money market rate, 2007	4.34%	Household saving rate, 2007	2.4%

Exchange rates

	end 2007		December 2007
C$ per $	0.99	Effective rates	2000 = 100
C$ per SDR	1.56	– nominal	142.10
C$ per €	1.46	– real	138.10

Trade

Principal exports		Principal imports	
	$bn fob		*$bn fob*
Machinery & equipment	83.5	Machinery & equipment	101.1
Other industrial goods	82.8	Industrial goods	74.0
Energy products	76.5	Motor vehicles & parts	70.3
Motor vehicles and parts	72.8	Consumer goods	45.9
Total incl. others	**389.4**	Total incl. others	**348.7**

Main export destinations		Main origins of imports	
	% of total		*% of total*
United States	76.4	United States	65.0
United Kingdom	3.2	Japan	2.9
Japan	2.2	United Kingdom	2.4
EU25	5.4	EU25	7.8

Balance of payments, reserves and aid, $bn

Visible exports fob	401.8	Overall balance	0.8
Visible imports fob	-356.6	Change in reserves	2.0
Trade balance	45.1	Level of reserves	
Invisibles inflows	113.7	end Dec.	35.1
Invisibles outflows	-137.4	No. months of import cover	0.9
Net transfers	-0.6	Official gold holdings, m oz	0.1
Current account balance	20.8	Aid given	3.68
– as % of GDP	1.6	– as % of GDP	0.29
Capital balance	-15.5		

Health and education

Health spending, % of GDP	9.7	Education spending, % of GDP	...
Doctors per 1,000 pop.	2.2	Enrolment, %: primary	100
Hospital beds per 1,000 pop.	3.6	secondary	117
Improved-water source access,		tertiary	62
% of pop.	100		

Society

No. of households	12.3m	Colour TVs per 100 households	99.0
Av. no. per household	2.6	Telephone lines per 100 pop.	64.5
Marriages per 1,000 pop.	4.5	Mobile telephone subscribers	
Divorces per 1,000 pop.	2.2	per 100 pop.	57.6
Cost of living, Dec. 2007		Computers per 100 pop.	87.6
New York = 100	101	Internet hosts per 1,000 pop.	144.7

a Including freshwater.

CHILE

Area	756,945 sq km	Capital	Santiago
Arable as % of total land	3	Currency	Chilean peso (Ps)

People

Population	16.5m	Life expectancy: men		75.5 yrs
Pop. per sq km	21.8	women		81.5 yrs
Av. ann. growth		Adult literacy		95.7%
in pop. 2010–15	0.90%	Fertility rate (per woman)		1.9
Pop. under 15	24.9%	Urban population		87.7%
Pop. over 60	11.6%			per 1,000 pop.
No. of men per 100 women	98	Crude birth rate		15.0
Human Development Index	86.7	Crude death rate		5.4

The economy

GDP	77,338bn pesos	GDP per head	$8,840
GDP	$146bn	GDP per head in purchasing	
Av. ann. growth in real		power parity (USA=100)	29.6
GDP 1996–2006	4.2%	Economic freedom index	79.8

Origins of GDP		**Components of GDP**	
	% of total		% of total
Agriculture	5.0	Private consumption	54.1
Industry, of which:	49.9	Public consumption	10.5
manufacturing	36.3	Investment	19.5
Services	45.1	Exports	45.7
		Imports	-30.8

Structure of employment

	% of total		% of labour force
Agriculture	13	Unemployed 2006	6.0
Industry	23	Av. ann. rate 1995–2006	7.0
Services	64		

Energy

	m TOE		
Total output	9.1	Net energy imports as %	
Total consumption	29.6	of energy use	69
Consumption per head,			
kg oil equivalent	1,815		

Inflation and finance

Consumer price		av. ann. increase 2001–06	
inflation 2007	4.4%	Narrow money (M1)	13.3%
Av. ann. inflation 2002–07	3.0%	Broad money	12.0%
Money market rate, 2007	5.36%		

Exchange rates

	end 2007		December 2007
Ps per $	495.82	Effective rates	2000 = 100
Ps per SDR	783.52	– nominal	96.80
Ps per €	729.90	– real	98.50

Trade

Principal exports		**Principal imports**	
	$bn fob		*$bn cif*
Copper	32.7	Intermediate goods	21.5
Fruit	2.4	Consumer goods	7.6
Paper products	1.9	Capital goods	6.1
Total incl. others	**58.5**	Total incl. others	**38.4**

Main export destinations		**Main origins of imports**	
	% of total		*% of total*
United States	15.3	United States	14.6
Japan	10.3	Argentina	11.7
China	8.4	Brazil	11.0
Netherlands	6.5	China	9.1
South Korea	5.8	South Korea	3.7

Balance of payments, reserves and debt, $bn

Visible exports fob	58.1	Change in reserves	2.5
Visible imports fob	-35.9	Level of reserves	
Trade balance	22.2	end Dec.	19.4
Invisibles inflows	10.8	No. months of import cover	3.5
Invisibles outflows	-31.2	Official gold holdings, m oz	0.0
Net transfers	3.4	Foreign debt	48.0
Current account balance	5.3	– as % of GDP	42
– as % of GDP	3.6	– as % of total exports	86
Capital balance	-4.8	Debt service ratio	20
Overall balance	2.0		

Health and education

Health spending, % of GDP	5.4	Education spending, % of GDP	3.2
Doctors per 1,000 pop.	1.3	Enrolment, %: primary	104
Hospital beds per 1,000 pop.	2.4	secondary	91
Improved-water source access,		tertiary	47
% of pop.	95		

Society

No. of households	4.5m	Colour TVs per 100 households	92.1
Av. no. per household	3.6	Telephone lines per 100 pop.	20.2
Marriages per 1,000 pop.	3.3	Mobile telephone subscribers	
Divorces per 1,000 pop.	0.5	per 100 pop.	75.6
Cost of living, Dec. 2007		Computers per 100 pop.	14.1
New York = 100	71	Internet hosts per 1,000 pop.	49.5

CHINA

Area	9,560,900 sq km	Capital	Beijing
Arable as % of total land	15	Currency	Yuan

People

Population	1,323.6m	Life expectancy:	men	71.3 yrs
Pop. per sq km	138.4		women	74.8 yrs
Av. ann. growth		Adult literacy		93.3%
in pop. 2010–15	0.54%	Fertility rate (per woman)		1.8
Pop. under 15	21.6%	Urban population		40.5%
Pop. over 60	11.0%		*per 1,000 pop.*	
No. of men per 100 women	107	Crude birth rate		12
Human Development Index	77.7	Crude death rate		7.1

The economy

GDP	Yuan21,087bn	GDP per head	$2,000
GDP	$2,645bn	GDP per head in purchasing	
Av. ann. growth in real		power parity (USA=100)	10.6
GDP 1996–2006	10.5%	Economic freedom index	52.8

Origins of GDP		**Components of GDP**	
	% of total		*% of total*
Agriculture	11.7	Private consumption	36.4
Industry, of which:	48.9	Public consumption	13.7
manufacturing	...	Investment	40.9
Services	39.3	Exports	39.7
		Imports	-31.9

Structure of employment

	% of total		*% of labour force*
Agriculture	43	Unemployed 2005	4.2
Industry	25	Av. ann. rate 1995–2005	3.5
Services	32		

Energy

	m TOE		
Total output	1,640.9	Net energy imports as %	
Total consumption	1,717.2	of energy use	4
Consumption per head,			
kg oil equivalent	1,316		

Inflation and finance

Consumer price		*av. ann. increase 2001–06*	
inflation 2007	4.8%	Narrow money (M1)	15.4%
Av. ann. inflation 2002–07	2.6%	Broad money	17.2%
Deposit rate, 2007	4.14%		

Exchange rates

	end 2007		*December 2007*
Yuan per $	7.30	Effective rates	*2000 = 100*
Yuan per SDR	11.54	– nominal	99.28
Yuan per €	10.75	– real	101.12

Trade

Principal exports		Principal imports	
	$bn fob		*$bn cif*
Office equipment	134.5	Electrical machinery	174.8
Telecoms equipment	123.6	Petroleum products	84.1
Electrical machinery	101.7	Professional &	
Apparel & clothing	95.4	scientific instruments	48.6
Misc. manufactures	55.5	Metalliferous ores & scrap	44.0
		Office equipment	40.7
Total incl. others	**968.9**	Total incl. others	**791.5**

Main export destinations		Main origins of imports	
	% of total		*% of total*
United States	21.0	Japan	14.6
Hong Kong	16.0	South Korea	11.3
Japan	9.5	Taiwan	11.0
South Korea	4.6	United States	7.5
Germany	4.2	Germany	4.8

Balance of payments, reserves and debt, $bn

Visible exports fob	969.7	Change in reserves	249.3
Visible imports fob	-751.9	Level of reserves	
Trade balance	217.7	end Dec.	1,080.8
Invisibles inflows	143.2	No. months of import cover	14.5
Invisibles outflows	-140.4	Official gold holdings, m oz	19.3
Net transfers	29.2	Foreign debt	322.8
Current account balance	249.9	– as % of GDP	14
– as % of GDP	9.4	– as % of total exports	35
Capital balance	10.0	Debt service ratio	3
Overall balance	246.9		

Health and education

Health spending, % of GDP	4.7	Education spending, % of GDP	...
Doctors per 1,000 pop.	1.5	Enrolment, %: primary	111
Hospital beds per 1,000 pop.	2.4	secondary	76
Improved-water source access,		tertiary	22
% of pop.	77		

Society

No. of households	376.5m	Colour TVs per 100 households	98.5
Av. no. per household	3.5	Telephone lines per 100 pop.	27.8
Marriages per 1,000 pop.	6.3	Mobile telephone subscribers	
Divorces per 1,000 pop.	1.4	per 100 pop.	34.8
Cost of living, Dec. 2007		Computers per 100 pop.	4.3
New York = 100	82	Internet hosts per 1,000 pop.	9.9

Note: Data excludes Special Administrative Regions ie, Hong Kong and Macau.

COLOMBIA

Area	1,141,748 sq km	Capital	Bogota
Arable as % of total land	2	Currency	Colombian peso (peso)

People

Population	46.3m	Life expectancy:	men	69.2 yrs
Pop. per sq km	40.6		women	76.6 yrs
Av. ann. growth		Adult literacy		92.8%
in pop. 2010–15	1.13%	Fertility rate (per woman)		2.1
Pop. under 15	30.3%	Urban population		77.4%
Pop. over 60	7.5%			per 1,000 pop.
No. of men per 100 women	97	Crude birth rate		20
Human Development Index	79.1	Crude death rate		5.5

The economy

GDP	321trn pesos	GDP per head	$3,310
GDP	$153bn	GDP per head in purchasing	
Av. ann. growth in real		power parity (USA=100)	14.5
GDP 1996–2006	3.3%	Economic freedom index	61.9

Origins of GDP		**Components of GDP**	
	% of total		% of total
Agriculture	12.0	Private consumption	61.3
Industry, of which:	35.4	Public consumption	17.8
manufacturing	16.4	Investment	22.8
Services	52.6	Exports	22.5
		Imports	-24.9

Structure of employment

	% of total		% of labour force
Agriculture	22	Unemployed 2006	12.7
Industry	19	Av. ann. rate 1995–2006	14.3
Services	59		

Energy

	m TOE		
Total output	48.2	Net energy imports as %	
Total consumption	28.6	of energy use	-178
Consumption per head,			
kg oil equivalent	636		

Inflation and finance

		av. ann. increase 2001–06	
Consumer price			
inflation 2007	5.4%	Narrow money (M1)	17.1%
Av. ann. inflation 2002–07	5.5%	Broad money	18.2%
Money market rate, 2007	8.66%		

Exchange rates

	end 2007		December 2007
Peso per $	1,988	Effective rates	2000 = 100
Peso per SDR	3,141	– nominal	106.50
Peso per €	2,926	– real	115.70

Trade

Principal exports		Principal imports	
	$bn fob		*$bn cif*
Oil	6.3	Intermediate goods &	
Coal	2.9	raw materials	11.5
Coffee	1.5	Capital goods	9.3
		Consumer goods	5.3
Total incl. others	**24.4**	Total	**26.2**

Main export destinations		Main origins of imports	
	% of total		*% of total*
United States	36.6	United States	28.2
Venezuela	10.6	Mexico	8.3
Ecuador	6.7	Brazil	6.5
Peru	3.6	China	6.3

Balance of payments, reserves and debt, $bn

Visible exports fob	25.2	Change in reserves	0.5
Visible imports fob	-24.9	Level of reserves	
Trade balance	0.3	end Dec.	15.4
Invisibles inflows	4.9	No. months of import cover	4.9
Invisibles outflows	-13.0	Official gold holdings, m oz	0.2
Net transfers	4.7	Foreign debt	39.7
Current account balance	-3.1	– as % of GDP	32
– as % of GDP	-2.0	– as % of total exports	143
Capital balance	2.8	Debt service ratio	31
Overall balance	0.0		

Health and education

Health spending, % of GDP	7.3	Education spending, % of GDP	4.7
Doctors per 1,000 pop.	1.6	Enrolment, %: primary	116
Hospital beds per 1,000 pop.	1.2	secondary	82
Improved-water source access,		tertiary	31
% of pop.	93		

Society

No. of households	12.4m	Colour TVs per 100 households	78.2
Av. no. per household	3.7	Telephone lines per 100 pop.	17.0
Marriages per 1,000 pop.	1.7	Mobile telephone subscribers	
Divorces per 1,000 pop.	0.2	per 100 pop.	64.3
Cost of living, Dec. 2007		Computers per 100 pop.	4.2
New York = 100	78	Internet hosts per 1,000 pop.	28.1

CÔTE D'IVOIRE

Area	322,463 sq km	Capital	Abidjan/Yamoussoukro
Arable as % of total land	11	Currency	CFA franc (CFAfr)

People

Population	18.5m	Life expectancy: men		47.5 yrs
Pop. per sq km	57.4	women		49.3 yrs
Av. ann. growth		Adult literacy		48.7%
in pop. 2010–15	1.84%	Fertility rate (per woman)		3.9
Pop. under 15	41.7%	Urban population		45.8%
Pop. over 60	5.1%			per 1,000 pop.
No. of men per 100 women	103	Crude birth rate		38
Human Development Index	43.2	Crude death rate		15.4

The economy

GDP	CFAfr9,177bn	GDP per head	$950
GDP	$17.6bn	GDP per head in purchasing	
Av. ann. growth in real		power parity (USA=100)	3.8
GDP 1997–2007	0.3%	Economic freedom index	54.9

Origins of GDP		Components of GDP	
	% of total		% of total
Agriculture	27.9	Private consumption	73.7
Industry, of which:	22.3	Public consumption	8.3
manufacturing	18.4	Investment	10.7
Services	49.8	Exports	49.6
		Imports	-42.4

Structure of employment

	% of total		% of labour force
Agriculture	...	Unemployed 2006	...
Industry	...	Av. ann. rate 1995–2006	...
Services	...		

Energy

	m TOE		
Total output	8.2	Net energy imports as %	
Total consumption	7.8	of energy use	-5
Consumption per head,			
kg oil equivalent	422		

Inflation and finance

			av. ann. change 2001–06
Consumer price			
inflation 2007	1.9%	Narrow money (M1)	3.2%
Av. ann. inflation 2002–07	2.6%	Broad money	4.5%
Money market rate, 2007	3.93%		

Exchange rates

	end 2007		December 2007
CFAfr per $	445.59	Effective rates	2000 = 100
CFAfr per SDR	704.15	– nominal	117.30
CFAfr per €	655.95	– real	117.10

Trade

Principal exports[a]	$bn fob	Principal imports[a]	$bn cif
Cocoa beans & products	2.1	Capital equipment	
Petroleum products	2.0	& raw materials	2.6
Timber	0.3	Fuel & lubricants	1.6
Coffee & products	0.1	Foodstuffs	0.9
Total incl. others	**7.7**	Total incl. others	**5.9**

Main export destinations	% of total	Main origins of imports	% of total
France	18.3	Nigeria	27.6
Netherlands	9.7	France	25.4
United States	9.1	China	4.3
Nigeria	7.2	Venezuela	2.8
Germany	4.2	Germany	2.5

Balance of payments, reserves and debt, $bn

Visible exports fob	8.2	Change in reserves	0.5
Visible imports fob	-5.0	Level of reserves	
Trade balance	3.2	end Dec.	1.8
Invisibles inflows	1.0	No. months of import cover	2.6
Invisibles outflows	-3.1	Official gold holdings, m oz	0.0
Net transfers	-0.5	Foreign debt	13.8
Current account balance	0.5	– as % of GDP	82
– as % of GDP	3.0	– as % of total exports	150
Capital balance	-0.4	Debt service ratio	1
Overall balance	0.2		

Health and education

Health spending, % of GDP	3.9	Education spending, % of GDP	...
Doctors per 1,000 pop.	0.1	Enrolment, %: primary	71
Hospital beds per 1,000 pop.	...	secondary	25
Improved-water source access,		tertiary	7
% of pop.	84		

Society

No. of households	3.6m	Colour TVs per 100 households	28.0
Av. no. per household	4.7	Telephone lines per 100 pop.	1.4
Marriages per 1,000 pop.	...	Mobile telephone subscribers	
Divorces per 1,000 pop.	...	per 100 pop.	22.0
Cost of living, Dec. 2007		Computers per 100 pop.	1.7
New York = 100	93	Internet hosts per 1,000 pop.	0.1

a 2005

CZECH REPUBLIC

Area	78,864 sq km	Capital	Prague
Arable as % of total land	39	Currency	Koruna (Kc)

People

Population	10.2m	Life expectancy:	men	73.4 yrs
Pop. per sq km	129.3		women	79.5 yrs
Av. ann. growth		Adult literacy		...
in pop. 2010–15	-0.09%	Fertility rate (per woman)		1.3
Pop. under 15	14.8%	Urban population		74.5%
Pop. over 60	19.9%			per 1,000 pop.
No. of men per 100 women	95	Crude birth rate		10
Human Development Index	89.1	Crude death rate		10.9

The economy

GDP	Kc3,232bn	GDP per head	$14,020
GDP	$143bn	GDP per head in purchasing	
Av. ann. growth in real		power parity (USA=100)	50.3
GDP 1997–2007	4.0%	Economic freedom index	68.5

Origins of GDP		Components of GDP	
	% of total		% of total
Agriculture	2.7	Private consumption	48.1
Industry, of which:	38.1	Public consumption	21.2
manufacturing	32.5	Investment	24.6
Services	59.3	Exports	75.8
		Imports	-72.7

Structure of employment

	% of total		% of labour force
Agriculture	4	Unemployed 2006	7.1
Industry	40	Av. ann. rate 1995–2006	6.9
Services	56		

Energy

	m TOE		
Total output	32.9	Net energy imports as %	
Total consumption	45.2	of energy use	27
Consumption per head,			
kg oil equivalent	4,417		

Inflation and finance

Consumer price		av. ann. increase 2001–06	
inflation 2007	2.9%	Narrow money (M1)	17.8%
Av. ann. inflation 2002–07	2.0%	Broad money	7.4%
Money market rate, 2007	4.11%	Household saving rate, 2007	1.2%

Exchange rates

	end 2007		December 2007
Kc per $	18.08	Effective rates	2000 = 100
Kc per SDR	28.57	– nominal	144.10
Kc per €	26.62	– real	147.29

Trade

Principal exports	$bn fob	Principal imports	$bn cif
Machinery & transport equipment	55.9	Machinery & transport equipment	44.6
Semi-manufactures	18.9	Semi-manufactures	18.9
Chemicals	5.7	Raw materials & fuels	11.7
Raw materials & fuels	5.5	Chemicals	9.4
Total incl. others	**95.1**	Total incl. others	**93.4**

Main export destinations	% of total	Main origins of imports	% of total
Germany	31.7	Germany	28.3
Slovakia	8.4	China	6.1
Russia	2.0	Russia	6.0
China	0.4	Slovakia	5.1
EU25	83.9	EU25	70.0

Balance of payments, reserves and debt, $bn

Visible exports fob	95.1	Change in reserves	1.9
Visible imports fob	-92.1	Level of reserves	
Trade balance	3.0	end Dec.	31.5
Invisibles inflows	18.7	No. months of import cover	3.2
Invisibles outflows	-25.4	Official gold holdings, m oz	0.4
Net transfers	-0.9	Foreign debt	58.0
Current account balance	-4.6	– as % of GDP	41
– as % of GDP	-3.2	– as % of total exports	51
Capital balance	5.5	Debt service ratio	11
Overall balance	0.1		

Health and education

Health spending, % of GDP	7.1	Education spending, % of GDP	4.4
Doctors per 1,000 pop.	3.6	Enrolment, %: primary	100
Hospital beds per 1,000 pop.	8.4	secondary	96
Improved-water source access, % of pop.	100	tertiary	50

Society

No. of households	3.8m	Colour TVs per 100 households	97.5
Av. no. per household	2.7	Telephone lines per 100 pop.	28.3
Marriages per 1,000 pop.	5.1	Mobile telephone subscribers	
Divorces per 1,000 pop.	3.3	per 100 pop.	121.5
Cost of living, Dec. 2007		Computers per 100 pop.	27.4
New York = 100	91	Internet hosts per 1,000 pop.	205.2

DENMARK

Area	43,075 sq km	Capital	Copenhagen
Arable as % of total land	53	Currency	Danish krone (DKr)

People

Population	5.4m	Life expectancy:	men	76.0 yrs
Pop. per sq km	125.4		women	80.6 yrs
Av. ann. growth		Adult literacy		...
in pop. 2010–15	0.14	Fertility rate (per woman)		1.8
Pop. under 15	18.8%	Urban population		85.5%
Pop. over 60	21.2%			*per 1,000 pop.*
No. of men per 100 women	98	Crude birth rate		12
Human Development Index	94.9	Crude death rate		10.3

The economy

GDP	DKr1,638bn	GDP per head	$50,990
GDP	$275bn	GDP per head in purchasing	
Av. ann. growth in real		power parity (USA=100)	81.2
GDP 1997–2007	2.2%	Economic freedom index	79.2

Origins of GDP		**Components of GDP**	
	% of total		*% of total*
Agriculture	1.6	Private consumption	49.5
Industry, of which:	26.0	Public consumption	24.7
manufacturing	...	Investment	23.5
Services	72.4	Exports	54.1
		Imports	-53.4

Structure of employment

	% of total		*% of labour force*
Agriculture	3	Unemployed 2006	4.1
Industry	24	Av. ann. rate 1995–2006	5.4
Services	73		

Energy

	m TOE		
Total output	31.3	Net energy imports as %	
Total consumption	19.6	of energy use	-60
Consumption per head,			
kg oil equivalent	3,621		

Inflation and finance

Consumer price		*av. ann. increase 2001–06*	
inflation 2007	1.7%	Narrow money (M1)	11.0%
Av. ann. inflation 2002–07	1.7%	Broad money	9.3%
Money market rate, 2007	4.33%	Household saving rate, 2007	3.2%

Exchange rates

	end 2007		*December 2007*
DKr per $	5.08	Effective rates	*2000 = 100*
DKr per SDR	8.02	– nominal	111.70
DKr per €	7.48	– real	110.90

Trade

Principal exports		**Principal imports**	
	$bn fob		*$bn cif*
Machinery & transport equip.	25.3	Machinery & transport equip.	31.4
Food, drinks & tobacco	16.5	Food, drinks & tobacco	9.8
Chemicals & related products	11.7	Chemicals & related products	9.6
Minerals, fuels & lubricants	10.3	Minerals, fuels & lubricants	4.8
Total incl. others	**85.3**	Total incl. others	**91.6**

Main export destinations		**Main origins of imports**	
	% of total		*% of total*
Germany	16.8	Germany	21.5
Sweden	14.2	Sweden	14.3
United Kingdom	9.0	United Kingdom	5.8
United States	6.7	China	5.2
Norway	5.7	Norway	4.6
EU25	69.6	EU25	72.6

Balance of payments, reserves and aid, $bn

Visible exports fob	90.6	Overall balance	-4.1
Visible imports fob	-87.9	Change in reserves	-2.9
Trade balance	2.7	Level of reserves	
Invisibles inflows	80.1	end Dec.	31.1
Invisibles outflows	-71.0	No. months of import cover	2.3
Net transfers	-4.5	Official gold holdings, m oz	2.1
Current account balance	7.3	Aid given	2.24
– as % of GDP	2.7	– as % of GDP	0.81
Capital balance	-4.5		

Health and education

Health spending, % of GDP	9.1	Education spending, % of GDP	8.3
Doctors per 1,000 pop.	3.6	Enrolment, %: primary	99
Hospital beds per 1,000 pop.	3.8	secondary	120
Improved-water source access,		tertiary	80
% of pop.	100		

Society

No. of households	2.5m	Colour TVs per 100 households	98.0
Av. no. per household	2.2	Telephone lines per 100 pop.	57.0
Marriages per 1,000 pop.	7.1	Mobile telephone subscribers	
Divorces per 1,000 pop.	3.0	per 100 pop.	107.0
Cost of living, Dec. 2007		Computers per 100 pop.	69.6
New York = 100	137	Internet hosts per 1,000 pop.	603.0

EGYPT

Area	1,000,250 sq km	Capital	Cairo
Arable as % of total land	3	Currency	Egyptian pound (£E)

People

Population	75.4m	Life expectancy: men	69.1 yrs
Pop. per sq km	75.4	women	73.6 yrs
Av. ann. growth		Adult literacy	72.0%
in pop. 2010–15	1.61%	Fertility rate (per woman)	2.7
Pop. under 15	33.3%	Urban population	42.3%
Pop. over 60	7.2%		per 1,000 pop.
No. of men per 100 women	100	Crude birth rate	27
Human Development Index	70.8	Crude death rate	5.6

The economy

GDP	£E618bn	GDP per head	$1,430
GDP	$107bn	GDP per head in purchasing	
Av. ann. growth in real		power parity (USA=100)	11.3
GDP 1997–2007	5.7%	Economic freedom index	59.2

Origins of GDP		**Components of GDP**a	
	% of total		% of total
Agriculture	14.1	Private consumption	70.4
Industry, of which:	38.4	Public consumption	11.5
manufacturing	15.5	Investment	21.2
Services	45.4	Exports	31.5
		Imports	-34.8

Structure of employment

	% of total		% of labour force
Agriculture	30	Unemployed 2005	11.2
Industry	20	Av. ann. rate 1995–2005	9.9
Services	50		

Energy

	m TOE		
Total output	76.0	Net energy imports as %	
Total consumption	61.3	of energy use	-24
Consumption per head,			
kg oil equivalent	841		

Inflation and finance

Consumer price		av. ann. increase 2001–06	
inflation 2007	9.3%	Narrow money (M1)	12.6%
Av. ann. inflation 2002–07	7.5%	Broad money	15.3%
Treasury bill rate, 2007	6.85%		

Exchange rates

	end 2007		December 2007
£E per $	5.55	Effective rates	2000 = 100
£E per SDR	8.65	– nominal	...
£E per €	8.18	– real	...

Trade

Principal exports[b]

	$bn fob
Petroleum & products	10.7
Finished goods incl. textiles	6.3
Semi-finished products	1.5
Iron & steel	0.7
Total incl. others	**20.5**

Principal imports[b]

	$bn fob
Intermediate goods	8.7
Investment goods	8.7
Consumer goods	4.2
Fuels	2.7
Total incl. others	**33.1**

Main export destinations

	% of total
Italy	12.2
United States	11.4
Spain	8.6
United Kingdom	5.6

Main origins of imports

	% of total
United States	11.4
China	8.2
Germany	6.4
Italy	5.4

Balance of payments, reserves and debt, $bn

Visible exports fob	20.5	Change in reserves	4.2
Visible imports fob	-29.0	Level of reserves	
Trade balance	-8.4	end Dec.	26.0
Invisibles inflows	18.7	No. months of import cover	7.4
Invisibles outflows	-13.4	Official gold holdings, m oz	2.4
Net transfers	5.7	Foreign debt	29.4
Current account balance	2.6	– as % of GDP	28
– as % of GDP	2.5	– as % of total exports	69
Capital balance	-0.5	Debt service ratio	5
Overall balance	2.9		

Health and education

Health spending, % of GDP	6.1	Education spending, % of GDP	4.2
Doctors per 1,000 pop.	0.6	Enrolment, %: primary	105
Hospital beds per 1,000 pop.	2.2	secondary	88
Improved-water source access,		tertiary	35
% of pop.	98		

Society

No. of households	16.4m	Colour TVs per 100 households	78.2
Av. no. per household	4.4	Telephone lines per 100 pop.	14.3
Marriages per 1,000 pop.	6.7	Mobile telephone subscribers	
Divorces per 1,000 pop.	0.8	per 100 pop.	23.9
Cost of living, Dec. 2007		Computers per 100 pop.	3.7
New York = 100	61	Internet hosts per 1,000 pop.	2.3

a Year ending June 30, 2007.
b Year ending June 30, 2006.

ESTONIA

Area	45,200 sq km	Capital	Tallinn
Arable as % of total land	14	Currency	Kroon (EEK)

People

Population	1.3m	Life expectancy: men	65.9 yrs
Pop. per sq km	28.8	women	76.8 yrs
Av. ann. growth		Adult literacy	99.8%
in pop. 2010–15	-0.33%	Fertility rate (per woman)	1.5
Pop. under 15	15.2%	Urban population	69.6%
Pop. over 60	21.6%		per 1,000 pop.
No. of men per 100 women	85	Crude birth rate	11
Human Development Index	86.0	Crude death rate	14.3

The economy

GDP	EEK205bn	GDP per head	$12,620
GDP	$16.4bn	GDP per head in purchasing	
Av. ann. growth in real		power parity (USA=100)	43.1
GDP 1997–2007	8.3%	Economic freedom index	77.8

Origins of GDP		**Components of GDP**	
	% of total		% of total
Agriculture	3.1	Private consumption	54.1
Industry, of which:	28.4	Public consumption	16.4
manufacturing	21.0	Investment	34.1
Services	68.5	Exports	79.4
		Imports	-90.7

Structure of employment

	% of total		% of labour force
Agriculture	5	Unemployed 2006	5.9
Industry	34	Av. ann. rate 1995–2006	10.0
Services	61		

Energy

	m TOE		
Total output	3.7	Net energy imports as %	
Total consumption	5.1	of energy use	27
Consumption per head,			
kg oil equivalent	3,786		

Inflation and finance

Consumer price		av. ann. increase 2001–06	
inflation 2007	6.6%	Narrow money (M1)	19.9%
Av. ann. inflation 2002–07	3.9%	Broad money	21.0%
Money market rate, 2007	4.87%		

Exchange rates

	end 2007		December 2007
EEK per $	10.64	Effective rates	2000 = 100
EEK per SDR	16.81	– nominal	...
EEK per €	15.66	– real	...

Trade

Principal exports		Principal imports	
	$bn fob		*$bn cif*
Machinery & equipment	1.6	Machinery & equipment	3.4
Wood & paper	1.1	Transport equipment	1.6
Food	0.7	Chemicals	1.5
Furniture	0.7	Metals	1.3
Clothing & footwear	0.6	Food	0.9
Total incl. others	**9.7**	Total incl. others	**16.7**

Main export destinations		Main origins of imports	
	% of total		*% of total*
Finland	18.2	Finland	18.3
Sweden	12.2	Russia	13.0
Latvia	9.1	Germany	12.4
Russia	7.8	Sweden	9.1
Germany	5.0	Lithuania	6.5

Balance of payments, reserves and debt, $bn

Visible exports fob	9.6	Change in reserves	0.8
Visible imports fob	-12.4	Level of reserves	
Trade balance	-2.7	end Dec.	2.8
Invisibles inflows	4.5	No. months of import cover	2.0
Invisibles outflows	-4.2	Official gold holdings, m oz	0.0
Net transfers	0.0	Foreign debt[a]	11.3
Current account balance	-2.4	– as % of GDP[a]	102
– as % of GDP	-14.9	– as % of total exports[a]	115
Capital balance	3.1	Debt service ratio[a]	14
Overall balance	0.6		

Health and education

Health spending, % of GDP	5.0	Education spending, % of GDP	5.1
Doctors per 1,000 pop.	3.4	Enrolment, %: primary	99
Hospital beds per 1,000 pop.	5.8	secondary	100
Improved-water source access,		tertiary	50
% of pop.	100		

Society

No. of households	0.6m	Colour TVs per 100 households	95.5
Av. no. per household	2.4	Telephone lines per 100 pop.	34.1
Marriages per 1,000 pop.	4.6	Mobile telephone subscribers	
Divorces per 1,000 pop.	3.1	per 100 pop.	125.2
Cost of living, Dec. 2007		Computers per 100 pop.	48.3
New York = 100	...	Internet hosts per 1,000 pop.	434.3

a 2005

FINLAND

Area	338,145 sq km	Capital	Helsinki
Arable as % of total land	7	Currency	Euro (€)

People

Population	5.3m	Life expectancy:	men	76.1 yrs
Pop. per sq km	15.7		women	82.4 yrs
Av. ann. growth		Adult literacy		...
in pop. 2010–15	0.23%	Fertility rate (per woman)		1.9
Pop. under 15	17.4%	Urban population		60.9%
Pop. over 60	21.4%			per 1,000 pop.
No. of men per 100 women	96	Crude birth rate		11
Human Development Index	95.2	Crude death rate		9.7

The economy

GDP	€168bn	GDP per head	$39,750
GDP	$211bn	GDP per head in purchasing	
Av. ann. growth in real		power parity (USA=100)	75.1
GDP 1997–2007	4.0%	Economic freedom index	74.8

Origins of GDP		**Components of GDP**	
	% of total		% of total
Agriculture	2.6	Private consumption	51.4
Industry, of which:	32.3	Public consumption	21.8
manuf., mining & utilities	...	Investment	19.3
Services	65.0	Exports	45.2
		Imports	-40.2

Structure of employment

	% of total		% of labour force
Agriculture	5	Unemployed 2006	7.7
Industry	26	Av. ann. rate 1995–2006	10.4
Services	69		

Energy

	m TOE		
Total output	16.6	Net energy imports as %	
Total consumption	35	of energy use	53
Consumption per head,			
kg oil equivalent	6,664		

Inflation and finance

Consumer price			av. ann. increase 2001–06
inflation 2007	2.5%	Euro area:	
Av. ann. inflation 2002–07	1.2%	Narrow money (M1)	10.5%
Money market rate, 2007	4.28%	Broad money	7.4%
		Household saving rate, 2007	-3.6%

Exchange rates

	end 2007		December 2007
€ per $	0.68	Effective rates	2000 = 100
€ per SDR	1.07	– nominal	115.70
		– real	107.40

Trade

Principal exports

	$bn fob
Machinery & transport equipment	33.4
Raw materials	4.8
Chemicals & related products	4.5
Mineral fuels & lubricants	4.2
Total incl. others	**77.3**

Principal imports

	$bn cif
Machinery & transport equipment	26.7
Minerals & fuels	10.9
Chemicals & related products	7.4
Raw materials	6.6
Total incl. others	**69.4**

Main export destinations

	% of total
Germany	11.3
Sweden	10.5
Russia	10.1
United Kingdom	6.5
United States	6.5
Netherlands	5.1

Main origins of imports

	% of total
Germany	15.5
Russia	13.9
Sweden	13.6
Netherlands	6.6
China	5.3
United Kingdom	4.6

Balance of payments, reserves and aid, $bn

Visible exports fob	77.5	Overall balance	-4.3
Visible imports fob	-66.4	Change in reserves	-3.8
Trade balance	11.2	Level of reserves	
Invisibles inflows	34.4	end Dec.	7.5
Invisibles outflows	-33.0	No. months of import cover	0.9
Net transfers	-1.7	Official gold holdings, m oz	1.6
Current account balance	10.9	Aid given	0.83
– as % of GDP	5.2	– as % of GDP	0.40
Capital balance	-15.4		

Health and education

Health spending, % of GDP	7.5	Education spending, % of GDP	6.4
Doctors per 1,000 pop.	3.3	Enrolment, %: primary	98
Hospital beds per 1,000 pop.	7.0	secondary	112
Improved-water source access,		tertiary	93
% of pop.	100		

Society

No. of households	2.4m	Colour TVs per 100 households	96.9
Av. no. per household	2.2	Telephone lines per 100 pop.	36.3
Marriages per 1,000 pop.	5.7	Mobile telephone subscribers	
Divorces per 1,000 pop.	2.5	per 100 pop.	107.8
Cost of living, Dec. 2007		Computers per 100 pop.	50.0
New York = 100	125	Internet hosts per 1,000 pop.	703.5

FRANCE

Area	543,965 sq km	Capital	Paris
Arable as % of total land	34	Currency	Euro (€)

People

Population	60.7m	Life expectancy: men	77.1 yrs
Pop. per sq km	111.6	women	84.1 yrs
Av. ann. growth		Adult literacy	...
in pop. 2010–15	0.39%	Fertility rate (per woman)	1.9
Pop. under 15	18.4%	Urban population	76.7%
Pop. over 60	20.8%		per 1,000 pop.
No. of men per 100 women	95	Crude birth rate	13
Human Development Index	95.2	Crude death rate	8.9

The economy

GDP	€1,792bn	GDP per head	$37,040
GDP	$2,248bn	GDP per head in purchasing	
Av. ann. growth in real		power parity (USA=100)	72.8
GDP 1997–2007	2.5%	Economic freedom index	65.4

Origins of GDP		**Components of GDP**	
	% of total		% of total
Agriculture	2.0	Private consumption	56.7
Industry, of which:	20.8	Public consumption	23.6
manufacturing	...	Investment	20.5
Services	77.2	Exports	26.9
		Imports	-28.3

Structure of employment

	% of total		% of labour force
Agriculture	2	Unemployed 2005	9.8
Industry	24	Av. ann. rate 1995–2005	10.6
Services	74		

Energy

	m TOE		
Total output	136.9	Net energy imports as %	
Total consumption	27.6	of energy use	50
Consumption per head,			
kg oil equivalent	4,534		

Inflation and finance

Consumer price		*av. ann. increase 2001–06*	
inflation 2007	1.5%	Euro area:	
Av. ann. inflation 2002–07	1.8%	Narrow money (M1)	10.5%
Deposit rate, households, 2007	3.69%	Broad money	7.4%
		Household saving rate, 2007	13.1%

Exchange rates

	end 2007		December 2007
€ per $	0.68	Effective rates	2000 = 100
€ per SDR	1.07	– nominal	113.60
		– real	110.80

Trade

Principal exports		Principal imports	
	$bn fob		*$bn cif*
Intermediate goods	148.8	Intermediate goods	157.9
Capital goods	118.3	Capital goods	110.8
Consumer goods	73.6	Consumer goods	84.0
Motor vehicles & other		Energy	80.0
transport equipment	65.1	Motor vehicles & other	
Food & drink	40.9	transport equipment	56.9
Total incl. others	**482.6**	Total incl. others	**535.3**

Main export destinations		Main origins of imports	
	% of total		*% of total*
Germany	14.5	Germany	16.2
Spain	9.9	Belgium-Luxembourg	8.7
Italy	9.1	Italy	8.5
United Kingdom	8.5	Spain	6.9
Belgium-Luxembourg	7.8	United Kingdom	6.8
EU25	68.3	EU25	64.6

Balance of payments, reserves and aid, $bn

Visible exports fob	483.1	Overall balance	11.8
Visible imports fob	-520.8	Change in reserves	23.9
Trade balance	-37.7	Level of reserves	
Invisibles inflows	304.6	end Dec.	98.2
Invisibles outflows	-267.7	No. months of import cover	1.5
Net transfers	-27.5	Official gold holdings, m oz	87.4
Current account balance	-28.3	Aid given	10.60
– as % of GDP	-1.3	– as % of GDP	0.47
Capital balance	87.8		

Health and education

Health spending, % of GDP	11.1	Education spending, % of GDP	5.7
Doctors per 1,000 pop.	3.4	Enrolment, %: primary	110
Hospital beds per 1,000 pop.	7.5	secondary	114
Improved-water source access,		tertiary	56
% of pop.	100		

Society

No. of households	25.5m	Colour TVs per 100 households	95.3
Av. no. per household	2.4	Telephone lines per 100 pop.	55.8
Marriages per 1,000 pop.	4.1	Mobile telephone subscribers	
Divorces per 1,000 pop.	2.2	per 100 pop.	85.1
Cost of living, Dec. 2007		Computers per 100 pop.	57.5
New York = 100	141	Internet hosts per 1,000 pop.	236.5

GERMANY

Area	357,868 sq km	Capital	Berlin
Arable as % of total land	34	Currency	Euro (€)

People

Population	82.7m	Life expectancy:	men	76.5 yrs
Pop. per sq km	231.0		women	82.1 yrs
Av. ann. growth		Adult literacy		...
in pop. 2010–15	-0.13%	Fertility rate (per woman)		1.4
Pop. under 15	14.4%	Urban population		88.5%
Pop. over 60	25.1%			per 1,000 pop.
No. of men per 100 women	96	Crude birth rate		8
Human Development Index	93.5	Crude death rate		10.7

The economy

GDP	€2,309bn	GDP per head	$35,030
GDP	$2,897bn	GDP per head in purchasing	
Av. ann. growth in real		power parity (USA=100)	73.5
GDP 1997–2007	1.7%	Economic freedom index	71.2

Origins of GDP		Components of GDP	
	% of total		% of total
Agriculture	0.9	Private consumption	58.5
Industry, of which:	29.4	Public consumption	18.3
manufacturing	...	Investment	18.0
Services	69.8	Exports	45.1
		Imports	-39.6

Structure of employment

	% of total		% of labour force
Agriculture	2	Unemployed 2006	10.3
Industry	30	Av. ann. rate 1995–2006	9.3
Services	68		

Energy

	m TOE		
Total output	134.5	Net energy imports as %	
Total consumption	344.7	of energy use	61
Consumption per head,			
kg oil equivalent	4,180		

Inflation and finance

Consumer price		av. ann. increase 2001–06	
inflation 2007	2.1%	Euro area:	
Av. ann. inflation 2002–07	1.7%	Narrow money (M1)	10.5%
Deposit rate, households, 2007	3.61%	Broad money	7.4%
		Household saving rate, 2007	11.1%

Exchange rates

	end 2007		December 2007
€ per $	0.68	Effective rates	2000 = 100
€ per SDR	1.07	– nominal	116.30
		– real	110.10

Trade

Principal exports		Principal imports	
	$bn fob		*$bn cif*
Road vehicles	210.4	Computer technology	105.0
Machinery	159.5	Fuels	97.1
Chemicals	148.0	Chemicals	86.9
Computer equipment	125.6	Road vehicles	73.8
		Machinery	52.9
Total incl. others	**1,121.7**	Total incl. others	**921.7**

Main export destinations		Main origins of imports	
	% of total		*% of total*
France	9.6	Netherlands	11.5
United States	8.5	France	8.6
United Kingdom	7.2	Belgium	7.5
Italy	6.7	United Kingdom	5.9
Netherlands	6.2	China	5.8
Belgium	5.5	Italy	5.4
Austria	5.4	United States	5.0

Balance of payments, reserves and aid, $bn

Visible exports fob	1,131.3	Overall balance	-3.7
Visible imports fob	-934.1	Change in reserves	10.0
Trade balance	197.2	Level of reserves	
Invisibles inflows	409.1	end Dec.	111.6
Invisibles outflows	-422.2	No. months of import cover	1.0
Net transfers	-33.3	Official gold holdings, m oz	110.0
Current account balance	150.8	Aid given	10.43
– as % of GDP	5.2	– as % of GDP	0.36
Capital balance	-179.5		

Health and education

Health spending, % of GDP	10.7	Education spending, % of GDP	4.6
Doctors per 1,000 pop.	3.4	Enrolment, %: primary	101
Hospital beds per 1,000 pop.	8.4	secondary	100
Improved-water source access,		tertiary	46
% of pop.	100		

Society

No. of households	39.4m	Colour TVs per 100 households	97.7
Av. no. per household	2.1	Telephone lines per 100 pop.	65.9
Marriages per 1,000 pop.	4.8	Mobile telephone subscribers	
Divorces per 1,000 pop.	2.8	per 100 pop.	103.6
Cost of living, Dec. 2007		Computers per 100 pop.	60.6
New York = 100	115	Internet hosts per 1,000 pop.	249.8

GREECE

Area	131,957 sq km	Capital	Athens
Arable as % of total land	20	Currency	Euro (€)

People

Population	11.1m	Life expectancy: men	77.1 yrs
Pop. per sq km	84.1	women	81.9 yrs
Av. ann. growth		Adult literacy	96.0%
in pop. 2010–15	0.10%	Fertility rate (per woman)	1.4
Pop. under 15	14.3%	Urban population	61.4%
Pop. over 60	23.3%		per 1,000 pop.
No. of men per 100 women	98	Crude birth rate	10
Human Development Index	92.6	Crude death rate	9.9

The economy

GDP	€246bn	GDP per head	$27,790
GDP	$308bn	GDP per head in purchasing	
Av. ann. growth in real		power parity (USA=100)	71.4
GDP 1996–2006	4.6%	Economic freedom index	60.1

Origins of GDP

	% of total
Agriculture	3.7
Industry, of which:	22.2
mining & manufacturing	9.7
Services	73.5

Components of GDP

	% of total
Private consumption	71.0
Public consumption	15.8
Investment	25.8
Exports	21.9
Imports	-34.5

Structure of employment

	% of total		% of labour force
Agriculture	12	Unemployed 2006	8.8
Industry	23	Av. ann. rate 1995–2006	10.2
Services	65		

Energy

	m TOE		
Total output	10.3	Net energy imports as %	
Total consumption	31	of energy use	67
Consumption per head,			
kg oil equivalent	2,790		

Inflation and finance

Consumer price		av. ann. increase 2001–06	
inflation 2007	2.9%	Euro area:	
Av. ann. inflation 2002–07	3.2%	Narrow money (M1)	10.5%
Treasury bill rate, 2007	4.45%	Broad money	7.4%

Exchange rates

	end 2007		December 2007
€ per $	0.68	Effective rates	2000 = 100
€ per SDR	1.07	– nominal	114.00
		– real	118.70

Trade

Principal exports		Principal imports	
	$bn fob		*$bn cif*
Food	3.6	Machinery & transport equip.	18.3
Machinery	2.7	Chemicals & plastics	8.8
Machinery & transport equip.	2.7	Food	6.6
Total incl. others	**20.8**	Total incl. others	**63.5**

Main export destinations		Main origins of imports	
	% of total		*% of total*
Germany	11.3	Germany	12.5
Italy	11.3	Italy	11.5
Bulgaria	6.4	Russia	7.1
United Kingdom	6.0	France	5.9
EU25	63.9	EU25	57.3

Balance of payments, reserves and debt, $bn

Visible exports fob	20.3	Overall balance	0.3
Visible imports fob	-64.6	Change in reserves	0.6
Trade balance	-44.3	Level of reserves	
Invisibles inflows	40.3	end Dec.	2.9
Invisibles outflows	-29.9	No. months of import cover	0.4
Net transfers	4.3	Official gold holdings, m oz	3.6
Current account balance	-29.6	Aid given	0.42
– as % of GDP	-9.6	– as % of GDP	0.14
Capital balance	29.5		

Health and education

Health spending, % of GDP	10.1	Education spending, % of GDP	4.4
Doctors per 1,000 pop.	5.0	Enrolment, %: primary	102
Hospital beds per 1,000 pop.	4.7	secondary	103
Improved-water source access,		tertiary	95
% of pop.	...		

Society

No. of households	3.9m	Colour TVs per 100 households	99.4
Av. no. per household	2.9	Telephone lines per 100 pop.	55.4
Marriages per 1,000 pop.	5.6	Mobile telephone subscribers	
Divorces per 1,000 pop.	1.2	per 100 pop.	98.6
Cost of living, Dec. 2007		Computers per 100 pop.	9.2
New York = 100	89	Internet hosts per 1,000 pop.	119.5

HONG KONG

Area	1,075 sq km	Capital	Victoria
Arable as % of total land	5	Currency	Hong Kong dollar (HK$)

People

Population	7.1m	Life expectancy: men	79.4 yrs
Pop. per sq km	6,604.7	women	85.1 yrs
Av. ann. growth		Adult literacy	...
in pop. 2010–15	0.86%	Fertility rate (per woman)	1.0
Pop. under 15	15.1%	Urban population	100.0%
Pop. over 60	15.4%		per 1,000 pop.
No. of men per 100 women	92	Crude birth rate	10
Human Development Index	93.7	Crude death rate	5.9

The economy

GDP	HK$1,474bn	GDP per head	$26,750
GDP	$190bn	GDP per head in purchasing	
Av. ann. growth in real		power parity (USA=100)	88.8
GDP 1996–2006	4.2%	Economic freedom index	90.3

Origins of GDP		Components of GDP	
	% of total		% of total
Agriculture	0.1	Private consumption	59.9
Industry, of which:	8.1	Public consumption	8.1
manufacturing	3.2	Investment	20.3
Services	91.7	Exports	207.4
		Imports	-196.7

Structure of employment

	% of total		% of labour force
Agriculture	0	Unemployed 2006	4.8
Industry	15	Av. ann. rate 1995–2006	5.1
Services	85		

Energy

	m TOE		
Total output	0.0	Net energy imports as %	
Total consumption	18.1	of energy use	100
Consumption per head,			
kg oil equivalent	2,653		

Inflation and finance

Consumer price		av. ann. increase 2001–06	
inflation 2007	2.0%	Narrow money (M1)	10.5%
Av. ann. inflation 2002–07	0.4%	Broad money	6.7%
Money market rate, 2007	1.88%		

Exchange rates

	end 2007		December 2007
HK$ per $	7.81	Effective rates	2000 = 100
HK$ per SDR	12.34	– nominal	...
HK$ per €	11.49	– real	...

Trade

Principal exports[a]	$bn fob	Principal imports	$bn cif
Clothing	6.7	Raw materials &	
Office machinery	2.5	semi-manufactures	127.8
Miscellaneous manufactures	2.0	Capital goods	99.1
Electrical machinery & aparatus	1.7	Consumer goods	89.9
Plastics	0.7	Fuel	9.5
		Food	8.4
Total incl. others	**17.5**	Total incl. others	**334.6**

Main export destinations	% of total	Main origins of imports	% of total
China	48.7	China	46.3
United States	13.7	Japan	10.0
Japan	4.5	Taiwan	7.2
Germany	3.0	Singapore	6.7

Balance of payments, reserves and debt, $bn

Visible exports fob	317.6	Change in reserves	8.9
Visible imports fob	-331.6	Level of reserves	
Trade balance	-14.0	end Dec.	133.2
Invisibles inflows	155.1	No. months of import cover	3.5
Invisibles outflows	-118.7	Official gold holdings, m oz	0.1
Net transfers	-2.9	Foreign debt[b]	72.3
Current account balance	20.2	– as % of GDP[b]	41
– as % of GDP	10.6	– as % of total exports[b]	17
Capital balance	-19.9	Debt service ratio[b]	2
Overall balance	6.0		

Health and education

Health spending, % of GDP	...	Education spending, % of GDP	3.9
Doctors per 1,000 pop.	1.4	Enrolment, %: primary	95
Hospital beds per 1,000 pop.	...	secondary	85
Improved-water source access,		tertiary	33
% of pop.	...		

Society

No. of households	2.2m	Colour TVs per 100 households	99.5
Av. no. per household	3.2	Telephone lines per 100 pop.	54.0
Marriages per 1,000 pop.	6.0	Mobile telephone subscribers	
Divorces per 1,000 pop.	2.4	per 100 pop.	132.7
Cost of living, Dec. 2007		Computers per 100 pop.	61.2
New York = 100	107	Internet hosts per 1,000 pop.	115.0

a Domestic, excluding re-exports.
b 2005
Note: Hong Kong became a Special Administrative Region of China on July 1 1997.

HUNGARY

Area	93,030 sq km	Capital	Budapest
Arable as % of total land	51	Currency	Forint (Ft)

People

Population	10.1m	Life expectancy:	men	69.2 yrs
Pop. per sq km	108.6		women	77.2 yrs
Av. ann. growth		Adult literacy		...
in pop. 2010–15	-0.32%	Fertility rate (per woman)		1.3
Pop. under 15	15.8%	Urban population		65.9%
Pop. over 60	20.8%			per 1,000 pop.
No. of men per 100 women	91	Crude birth rate		10
Human Development Index	87.4	Crude death rate		13.2

The economy

GDP	Ft23,757bn	GDP per head	$11,180
GDP	$113bn	GDP per head in purchasing	
Av. ann. growth in real		power parity (USA=100)	41.6
GDP 1996–2006	4.6%	Economic freedom index	67.2

Origins of GDP		Components of GDP	
	% of total		% of total
Agriculture	3.3	Private consumption	66.3
Industry, of which:	30.9	Public consumption	10.2
manufacturing	23.3	Investment	21.7
Services	65.8	Exports	77.8
		Imports	-77.3

Structure of employment

	% of total		% of labour force
Agriculture	5	Unemployed 2006	7.5
Industry	32	Av. ann. rate 1995–2006	7.3
Services	63		

Energy

	m TOE		
Total output	10.3	Net energy imports as %	
Total consumption	27.8	of energy use	63
Consumption per head,			
kg oil equivalent	2,752		

Inflation and finance

Consumer price		av. ann. increase 2001–06	
inflation 2007	7.9%	Narrow money (M1)	16.0%
Av. ann. inflation 2002–07	5.3%	Broad money	12.6%
Treasury bill rate, 2007	7.67%		

Exchange rates

	end 2007		December 2007
Ft per $	172.61	Effective rates	2000 = 100
Ft per SDR	272.77	– nominal	112.00
Ft per €	254.10	– real	145.00

Trade

Principal exports		**Principal imports**	
	$bn fob		*$bn cif*
Machinery & equipment	45.8	Machinery & equipment	38.2
Other manufactures	20.2	Other manufactures	24.8
Food, drink & tobacco	4.0	Fuels	8.4
Raw materials	1.4	Food, drink & tobacco	3.0
Total incl. others	**74.1**	Total incl. others	**77.1**

Main export destinations		**Main origins of imports**	
	% of total		*% of total*
Germany	29.4	Germany	27.1
Italy	5.6	Russia	8.3
Austria	4.5	China	7.0
France	4.9	Austria	6.2

Balance of payments, reserves and debt, $bn

Visible exports fob	74.3	Change in reserves	3.0
Visible imports fob	-75.5	Level of reserves	
Trade balance	-1.1	end Dec.	21.6
Invisibles inflows	18.9	No. months of import cover	2.6
Invisibles outflows	-25.6	Official gold holdings, m oz	0.1
Net transfers	0.4	Foreign debt	107.7
Current account balance	-7.4	– as % of GDP	100
– as % of GDP	-6.6	– as % of total exports	127
Capital balance	12.8	Debt service ratio	33
Overall balance	1.1		

Health and education

Health spending, % of GDP	7.8	Education spending, % of GDP	5.5
Doctors per 1,000 pop.	3.0	Enrolment, %: primary	97
Hospital beds per 1,000 pop.	7.9	secondary	96
Improved-water source access,		tertiary	69
% of pop.	99		

Society

No. of households	3.7m	Colour TVs per 100 households	98.7
Av. no. per household	2.7	Telephone lines per 100 pop.	33.4
Marriages per 1,000 pop.	4.3	Mobile telephone subscribers	
Divorces per 1,000 pop.	2.4	per 100 pop.	99.0
Cost of living, Dec. 2007		Computers per 100 pop.	14.9
New York = 100	78	Internet hosts per 1,000 pop.	167.3

INDIA

Area	3,287,263 sq km	Capital	New Delhi
Arable as % of total land	54	Currency	Indian rupee (Rs)

People

Population	1,119.5m	Life expectancy: men		63.2 yrs
Pop. per sq km	340.6		women	66.4 yrs
Av. ann. growth		Adult literacy		66.0%
in pop. 2010–15	1.31%	Fertility rate (per woman)		2.6
Pop. under 15	33.0%	Urban population		28.7%
Pop. over 60	7.1%			per 1,000 pop.
No. of men per 100 women	107	Crude birth rate		24
Human Development Index	61.9	Crude death rate		8.2

The economy

GDP	Rs41,257bn	GDP per head	$810
GDP	$912bn	GDP per head in purchasing	
Av. ann. growth in real		power parity (USA=100)	5.6
GDP 1997–2007	7.8%	Economic freedom index	54.2

Origins of GDP[a]		Components of GDP[a]	
	% of total		% of total
Agriculture	18.3	Private consumption	55.8
Industry, of which:	28.3	Public consumption	10.3
manufacturing	16.1	Investment	32.5
Services	52.4	Exports	22.1
		Imports	-25.1

Structure of employment

	% of total		% of labour force
Agriculture	...	Unemployed 2004	5.0
Industry	...	Av. ann. rate 1995–2004	3.3
Services	...		

Energy

	m TOE		
Total output	419	Net energy imports as %	
Total consumption	537.3	of energy use	22
Consumption per head,			
kg oil equivalent	491		

Inflation and finance

		av. ann. increase 2001–06	
Consumer price			
inflation 2007	6.4%	Narrow money (M1)	17.5%
Av. ann. inflation 2002–07	4.8%	Broad money	16.7%
Lending rate, 2007	13.02%		

Exchange rates

	end 2007		December 2007
Rs per $	39.42	Effective rates	2000 = 100
Rs per SDR	62.29	– nominal	...
Rs per €	56.03	– real	...

Trade

Principal exports[a]		Principal imports[a]	
	$bn fob		*$bn cif*
Engineering goods	26.2	Petroleum & products	57.1
Petroleum & products	18.6	Capital goods	32.7
Textiles	16.1	Electronic goods	15.9
Gems & jewellery	15.6	Gold & silver	14.6
Agricultural goods	11.2	Gems	7.8
Total incl. others	**126.3**	Total incl. others	**190.6**

Main export destinations		Main origins of imports	
	% of total		*% of total*
United States	19.1	China	7.3
China	9.4	United States	6.5
United Arab Emirates	8.4	Belgium	5.2
United Kingdom	4.9	Singapore	4.8

Balance of payments, reserves and debt, $bn

Visible exports fob	123.6	Change in reserves	40.2
Visible imports fob	-166.7	Level of reserves	
Trade balance	-43.1	end Dec.	178.1
Invisibles inflows	83.1	No. months of import cover	8.8
Invisibles outflows	-75.6	Official gold holdings, m oz	11.5
Net transfers	26.1	Foreign debt	153.1
Current account balance	-9.4	– as % of GDP	15
– as % of GDP	-1.0	– as % of total exports	63
Capital balance	37.8	Debt service ratio	8
Overall balance	23.7		

Health and education

Health spending, % of GDP	5.0	Education spending, % of GDP	3.2
Doctors per 1,000 pop.	0.6	Enrolment, %: primary	112
Hospital beds per 1,000 pop.	0.9	secondary	54
Improved-water source access,		tertiary	11
% of pop.	86		

Society

No. of households	209.9m	Colour TVs per 100 households	73.0
Av. no. per household	5.4	Telephone lines per 100 pop.	3.6
Marriages per 1,000 pop.	...	Mobile telephone subscribers	
Divorces per 1,000 pop.	...	per 100 pop.	14.8
Cost of living, Dec. 2007		Computers per 100 pop.	1.6
New York = 100	51	Internet hosts per 1,000 pop.	2.3

a Year ending March 31, 2007.

INDONESIA

Area	1,904,443 sq km	Capital	Jakarta
Arable as % of total land	13	Currency	Rupiah (Rp)

People

Population	225.5m	Life expectancy:	men	68.7 yrs
Pop. per sq km	118.4		women	72.7 yrs
Av. ann. growth		Adult literacy		91.4%
in pop. 2010–15	0.98%	Fertility rate (per woman)		2.0
Pop. under 15	28.4%	Urban population		47.9%
Pop. over 60	8.3%			per 1,000 pop.
No. of men per 100 women	100	Crude birth rate		21
Human Development Index	72.8	Crude death rate		6.3

The economy

GDP	Rp3,338trn	GDP per head	$1,620
GDP	$365bn	GDP per head in purchasing	
Av. ann. growth in real		power parity (USA=100)	7.9
GDP 1996–2006	3.0%	Economic freedom index	53.9

Origins of GDP		**Components of GDP**	
	% of total		% of total
Agriculture	13.0	Private consumption	62.7
Industry, of which:	46.9	Public consumption	8.6
manufacturing	27.5	Investment	24.1
Services	40.1	Exports	31.0
		Imports	-25.6

Structure of employment

	% of total		% of labour force
Agriculture	42	Unemployed 2006	10.5
Industry	19	Av. ann. rate 1995–2006	7.3
Services	39		

Energy

	m TOE		
Total output	263.4	Net energy imports as %	
Total consumption	179.5	of energy use	-47
Consumption per head,			
kg oil equivalent	814		

Inflation and finance

Consumer price		av. ann. increase 2001–06	
inflation 2007	6.4%	Narrow money (M1)	15.3%
Av. ann. inflation 2002–07	8.5%	Broad money	10.5%
Money market rate, 2007	6.02%		

Exchange rates

	end 2007		December 2007
Rp per $	9,419	Effective rates	2000 = 100
Rp per SDR	14,884	– nominal	...
Rp per €	13,866	– real	...

Trade

Principal exports		Principal imports	
	$bn fob		*$bn cif*
Petroleum & products	10.9	Raw materials	47.0
Natural gas	10.0	Capital goods	9.3
Garments & textiles	9.4	Consumer goods	4.5
Total incl. others	**100.6**	Total incl. others	**60.8**

Main export destinations		Main origins of imports	
	% of total		*% of total*
Japan	21.9	Singapore	45.0
United States	13.0	China	17.1
Singapore	13.3	Japan	13.3
South Korea	8.9	Malaysia	7.4

Balance of payments, reserves and debt, $bn

Visible exports fob	103.5	Change in reserves	7.9
Visible imports fob	-73.9	Level of reserves	
Trade balance	29.6	end Dec.	42.6
Invisibles inflows	14.1	No. months of import cover	4.5
Invisibles outflows	-38.7	Official gold holdings, m oz	2.4
Net transfers	4.9	Foreign debt	131.0
Current account balance	9.9	– as % of GDP	45
– as % of GDP	2.7	– as % of total exports	122
Capital balance	-1.0	Debt service ratio	17
Overall balance	11.4		

Health and education

Health spending, % of GDP	2.1	Education spending, % of GDP	3.6
Doctors per 1,000 pop.	0.1	Enrolment, %: primary	114
Hospital beds per 1,000 pop.	0.6	secondary	64
Improved-water source access,		tertiary	17
% of pop.	77		

Society

No. of households	59.6m	Colour TVs per 100 households	77.0
Av. no. per household	3.8	Telephone lines per 100 pop.	6.6
Marriages per 1,000 pop.	7.2	Mobile telephone subscribers	
Divorces per 1,000 pop.	0.8	per 100 pop.	28.3
Cost of living, Dec. 2007		Computers per 100 pop.	1.5
New York = 100	76	Internet hosts per 1,000 pop.	3.1

IRAN

Area	1,648,000 sq km	Capital	Tehran
Arable as % of total land	10	Currency	Rial (IR)

People

Population	70.3m	Life expectancy: men		69.4 yrs
Pop. per sq km	42.7	women		72.6 yrs
Av. ann. growth		Adult literacy		84.7%
in pop. 2010–15	1.33%	Fertility rate (per woman)		2.0
Pop. under 15	28.8%	Urban population		68.1%
Pop. over 60	6.4%			per 1,000 pop.
No. of men per 100 women	103	Crude birth rate		18
Human Development Index	75.9	Crude death rate		5.4

The economy

GDP	IR1,998trn	GDP per head	$3,100
GDP	$218bn	GDP per head in purchasing	
Av. ann. growth in real		power parity (USA=100)	22.5
GDP 1996–2006	5.5%	Economic freedom index	44.0

Origins of GDP		**Components of GDP**	
	% of total		% of total
Agriculture	10.4	Private consumption	45.2
Industry, of which:	40.8	Public consumption	14.2
manufacturing	11.6	Investment	26.4
Services	48.8	Exports	28.1
		Imports	-23.0

Structure of employment

	% of total		% of labour force
Agriculture	25	Unemployed 2005	11.5
Industry	30	Av. ann. rate 2000–2005	12.9
Services	45		

Energy

	m TOE		
Total output	303.8	Net energy imports as %	
Total consumption	162.5	of energy use	-87
Consumption per head,			
kg oil equivalent	2,352		

Inflation and finance

Consumer price		av. ann. increase 2001–06	
inflation 2007	17.2%	Narrow money (M1)	21.2%
Av. ann. inflation 2002–07	14.7%	Broad money	24.9%
Deposit rate, 2006	11.56%		

Exchange rates

	end 2007		December 2007
IR per $	9,282	Effective rates	2000 = 100
IR per SDR	14,668	– nominal	68.19
IR per €	13,664	– real	145.80

Trade

Principal exports		Principal imports	
	$bn fob		*$bn cif*
Oil & gas	62.5	Industrial goods	22.2
Chemicals & petrochemicals	2.9	Capital goods	18.2
Fruits	1.6	Consumer goods	8.9
Total incl. others	**75.5**	Total incl. others	**49.3**

Main export destinations		Main origins of imports	
	% of total		*% of total*
Japan	14.1	Germany	12.1
China	12.9	China	10.6
Turkey	7.3	United Arab Emirates	9.4
Italy	6.3	South Korea	6.2
South Korea	5.7	France	5.6

Balance of payments[a], reserves and debt, $bn

Visible exports fob	75.5	Change in reserves	...
Visible imports fob	-49.3	Level of reserves	
Trade balance	26.2	end Dec.	...
Net invisibles	-6.3	No. months of import cover	...
Net transfers	0.7	Official gold holdings, m oz	...
Current account balance	20.7	Foreign debt	20.1
– as % of GDP	9.5	– as % of GDP	10
Capital balance	-4.6	– as % of total exports	27
		Debt service ratio[b]	4

Health and education

Health spending, % of GDP	7.8	Education spending, % of GDP	5.1
Doctors per 1,000 pop.	0.9	Enrolment, %: primary	118
Hospital beds per 1,000 pop.	1.7	secondary	81
Improved-water source access,		tertiary	27
% of pop.	94		

Society

No. of households	13.1m	Colour TVs per 100 households	...
Av. no. per household	5.3	Telephone lines per 100 pop.	31.2
Marriages per 1,000 pop.	11.5	Mobile telephone subscribers	
Divorces per 1,000 pop.	1.2	per 100 pop.	21.8
Cost of living, Dec. 2007		Computers per 100 pop.	10.6
New York = 100	38	Internet hosts per 1,000 pop.	...

a Iranian year ending March 20, 2007.
b 2005

IRELAND

Area	70,282 sq km	Capital	Dublin
Arable as % of total land	18	Currency	Euro (€)

People

Population	4.2m	Life expectancy:	men	76.5 yrs
Pop. per sq km	59.8		women	81.4 yrs
Av. ann. growth		Adult literacy		...
in pop. 2010–15	1.20%	Fertility rate (per woman)		1.9
Pop. under 15	20.7%	Urban population		60.4%
Pop. over 60	15.3%			per 1,000 pop.
No. of men per 100 women	100	Crude birth rate		15
Human Development Index	95.9	Crude death rate		7.0

The economy

GDP	€175bn	GDP per head	$52,410
GDP	$220bn	GDP per head in purchasing	
Av. ann. growth in real		power parity (USA=100)	91.6
GDP 1996–2006	7.4%	Economic freedom index	82.4

Origins of GDP

	% of total
Agriculture	2.6
Industry, of which:	36.9
manufacturing	...
Services	57.3

Components of GDP

	% of total
Private consumption	47.2
Public consumption	14.3
Investment	26.3
Exports	80.0
Imports	-69.3

Structure of employment

	% of total		% of labour force
Agriculture	6	Unemployed 2006	4.3
Industry	28	Av. ann. rate 1995–2006	6.5
Services	66		

Energy

	m TOE		
Total output	1.7	Net energy imports as %	
Total consumption	15.3	of energy use	89
Consumption per head,			
kg oil equivalent	3,676		

Inflation and finance

Consumer price			av. ann. increase 2001–06
inflation 2007	4.9%	Euro area:	
Av. ann. inflation 2002–07	3.4%	Narrow money (M1)	10.5%
Deposit rate, households, 2007	3.81%	Broad money	7.4%

Exchange rates

	end 2007		December 2007
€ per $	0.68	Effective rates	2000 = 100
€ per SDR	1.07	– nominal	122.40
		– real	137.42

Trade

Principal exports		Principal imports	
	$bn fob		*$bn cif*
Chemicals	52.8	Machinery & transport	
Machinery & transport		equipment	29.9
equipment	29.2	Chemicals	10.2
Food; drink & tobacco	10.7	Food, drink & tobacco	6.2
Raw materials	2.2	Minerals, fuels & lubricants	5.3
Total incl. others	**109.0**	Total incl. others	**76.5**

Main export destinations		Main origins of imports	
	% of total		*% of total*
United States	19.3	United Kingdom	35.5
United Kingdom	17.8	United States	11.0
Belgium	16.3	Germany	9.1
Germany	7.7	Netherlands	4.3
France	5.8	France	3.5
Italy	4.2	China	2.9

Balance of payments, reserves and aid, $bn

Visible exports fob	104.7	Overall balance	-0.1
Visible imports fob	-72.8	Change in reserves	-0.0
Trade balance	31.9	Level of reserves	
Invisibles inflows	144.5	end Dec.	0.8
Invisibles outflows	-185.0	No. months of import cover	0.0
Net transfers	-0.5	Official gold holdings, m oz	0.2
Current account balance	-9.1	Aid given	1.02
– as % of GDP	-4.1	– as % of GDP	0.46
Capital balance	11.0		

Health and education

Health spending, % of GDP	8.2	Education spending, % of GDP	4.7
Doctors per 1,000 pop.	3.0	Enrolment, %: primary	104
Hospital beds per 1,000 pop.	5.7	secondary	112
Improved-water source access,		tertiary	59
% of pop.	...		

Society

No. of households	1.4m	Colour TVs per 100 households	99.5
Av. no. per household	3.0	Telephone lines per 100 pop.	49.9
Marriages per 1,000 pop.	5.1	Mobile telephone subscribers	
Divorces per 1,000 pop.	0.8	per 100 pop.	112.6
Cost of living, Dec. 2007		Computers per 100 pop.	52.8
New York = 100	115	Internet hosts per 1,000 pop.	297.1

ISRAEL

Area	20,770 sq km	Capital	Jerusalem[a]
Arable as % of total land	15	Currency	New Shekel (NIS)

People

Population	6.8m	Life expectancy: men		78.6 yrs
Pop. per sq km	327.4		women	82.8 yrs
Av. ann. growth		Adult literacy		97.1%
in pop. 2010–15	1.39%	Fertility rate (per woman)		2.6
Pop. under 15	27.9%	Urban population		91.7%
Pop. over 60	13.2%			per 1,000 pop.
No. of men per 100 women	98	Crude birth rate		21
Human Development Index	93.2	Crude death rate		5.5

The economy

GDP	NIS626bn	GDP per head	$20,660
GDP	$140bn	GDP per head in purchasing	
Av. ann. growth in real		power parity (USA=100)	54.8
GDP 1997–2007	4.2%	Economic freedom index	66.1

Origins of GDP		**Components of GDP**	
	% of total		% of total
Agriculture	2.8	Private consumption	54.3
Industry, of which:	32.0	Public consumption	26.0
manufacturing	22.1	Investment	17.1
Services	63.7	Exports	44.3
		Imports	-43.5

Structure of employment

	% of total		% of labour force
Agriculture	2	Unemployed 2006	8.4
Industry	22	Av. ann. rate 1995–2006	8.7
Services	76		

Energy

	m TOE		
Total output	2.1	Net energy imports as %	
Total consumption	19.5	of energy use	89
Consumption per head,			
kg oil equivalent	2,816		

Inflation and finance

Consumer price		*av. ann. increase 2001–06*	
inflation 2007	0.5%	Narrow money (M1)	11.4%
Av. ann. inflation 2002–07	0.8%	Broad money	3.0%
Interbank rate, 2007	4.02%		

Exchange rates

	end 2007		December 2007
NIS per $	3.85	Effective rates	2000 = 100
NIS per SDR	6.08	– nominal	86.67
NIS per €	5.67	– real	81.19

Trade

Principal exports		Principal imports	
	$bn fob		*$bn fob*
Diamonds	14.5	Diamonds	10.0
Chemicals	9.5	Fuel	8.9
Communications, medical &		Machinery & equipment	6.5
scientific equipment	8.0	Chemicals	3.6
Electronics	2.5		
Total incl. others	**45.9**	Total incl. others	**56.1**

Main export destinations		Main origins of imports	
	% of total		*% of total*
United States	45.0	United States	12.5
Belgium	7.6	Belgium	8.3
Hong Kong	6.9	Germany	6.8
Germany	4.4	Switzerland	5.9
United Kingdom	4.1	United Kingdom	5.2

Balance of payments, reserves and debt, $bn

Visible exports fob	43.7	Change in reserves	1.1
Visible imports fob	-47.0	Level of reserves	
Trade balance	-3.2	end Dec.	29.2
Invisibles inflows	27.2	No. months of import cover	5.0
Invisibles outflows	-23.4	Official gold holdings, m oz	0.0
Net transfers	7.5	Foreign debt	87.0
Current account balance	8.0	– as % of GDP	61
– as % of GDP	5.7	– as % of total exports	118
Capital balance	-12.3	Debt service ratio	12
Overall balance	-4.4		

Health and education

Health spending, % of GDP	7.9	Education spending, % of GDP	6.9
Doctors per 1,000 pop.	3.8	Enrolment, %: primary	110
Hospital beds per 1,000 pop.	6.3	secondary	92
Improved-water source access,		tertiary	58
% of pop.	100		

Society

No. of households	2.1m	Colour TVs per 100 households	94.1
Av. no. per household	3.4	Telephone lines per 100 pop.	43.9
Marriages per 1,000 pop.	5.1	Mobile telephone subscribers	
Divorces per 1,000 pop.	1.6	per 100 pop.	122.7
Cost of living, Dec. 2007		Computers per 100 pop.	122.1
New York = 100	93	Internet hosts per 1,000 pop.	205.6

a Sovereignty over the city is disputed.

ITALY

Area	301,245 sq km	Capital	Rome
Arable as % of total land	26	Currency	Euro (€)

People

Population	58.1m	Life expectancy:	men	77.5 yrs
Pop. per sq km	192.9		women	83.5 yrs
Av. ann. growth		Adult literacy		98.4%
in pop. 2010–15	-0.01%	Fertility rate (per woman)		1.4
Pop. under 15	14.0%	Urban population		67.5%
Pop. over 60	25.3%			per 1,000 pop.
No. of men per 100 women	94	Crude birth rate		10
Human Development Index	94.1	Crude death rate		10.5

The economy

GDP	€1,475bn	GDP per head	$31,860
GDP	$1,851bn	GDP per head in purchasing	
Av. ann. growth in real		power parity (USA=100)	66.1
GDP 1997–2007	1.6%	Economic freedom index	62.5

Origins of GDP		Components of GDP	
	% of total		% of total
Agriculture	1.9	Private consumption	58.8
Industry, of which:	28.9	Public consumption	20.6
manufacturing	...	Investment	21.0
Services	69.2	Exports	27.8
		Imports	-28.6

Structure of employment

	% of total		% of labour force
Agriculture	4	Unemployed 2006	6.8
Industry	31	Av. ann. rate 1995–2006	9.8
Services	65		

Energy

	m TOE		
Total output	27.6	Net energy imports as %	
Total consumption	185.2	of energy use	85
Consumption per head,			
kg oil equivalent	3,160		

Inflation and finance

Consumer price		av. ann. increase 2001–06	
inflation 2007	1.8%	Euro area:	
Av. ann. inflation 2002–07	2.2%	Narrow money (M1)	10.5%
Money market rate, 2007	4.29%	Broad money	7.4%
		Household saving rate, 2007	6.8%

Exchange rates

	end 2007		December 2007
€ per $	0.68	Effective rates	2000 = 100
€ per SDR	1.07	– nominal	115.90
		– real	113.70

Trade

Principal exports		Principal imports	
	$bn fob		*$bn cif*
Engineering products	144.6	Engineering products	90.9
Transport equipment	55.9	Energy products	70.3
Chemicals	46.6	Chemicals	64.6
Textiles & clothing	38.0	Transport equipment	64.5
Food, drink & tobacco	25.8	Food, drink & tobacco	31.4
Total incl. others	**417.4**	Total incl. others	**441.7**

Main export destinations		Main origins of imports	
	% of total		*% of total*
Germany	12.9	Germany	16.9
France	11.4	France	9.0
Spain	7.4	Netherlands	5.5
United States	6.8	Belgium	4.3
United Kingdom	5.8	United States	3.0
EU25	60.1	EU25	57.0

Balance of payments, reserves and aid, $bn

Visible exports fob	417.1	Overall balance	-0.6
Visible imports fob	-428.7	Change in reserves	9.8
Trade balance	-11.7	Level of reserves	
Invisibles inflows	170.9	end Dec.	75.8
Invisibles outflows	-189.9	No. months of import cover	1.5
Net transfers	-16.7	Official gold holdings, m oz	78.8
Current account balance	-47.3	Aid given	3.64
– as % of GDP	-2.6	– as % of GDP	0.20
Capital balance	46.0		

Health and education

Health spending, % of GDP	8.9	Education spending, % of GDP	4.5
Doctors per 1,000 pop.	3.7	Enrolment, %: primary	103
Hospital beds per 1,000 pop.	4.0	secondary	100
Improved-water source access,		tertiary	67
% of pop.	...		

Society

No. of households	22.6m	Colour TVs per 100 households	96.6
Av. no. per household	2.6	Telephone lines per 100 pop.	46.3
Marriages per 1,000 pop.	4.5	Mobile telephone subscribers	
Divorces per 1,000 pop.	0.8	per 100 pop.	135.1
Cost of living, Dec. 2007		Computers per 100 pop.	36.7
New York = 100	105	Internet hosts per 1,000 pop.	288.0

JAPAN

Area	377,727 sq km	Capital	Tokyo
Arable as % of total land	12	Currency	Yen (¥)

People

Population	128.2m	Life expectancy:	men	79.0 yrs
Pop. per sq km	339.4		women	86.1 yrs
Av. ann. growth		Adult literacy		...
in pop. 2010–15	-0.18%	Fertility rate (per woman)		1.3
Pop. under 15	13.9%	Urban population		65.7%
Pop. over 60	26.4%		*per 1,000 pop.*	
No. of men per 100 women	95	Crude birth rate		9
Human Development Index	95.3	Crude death rate		9.0

The economy

GDP	¥508trn	GDP per head	$34,080
GDP	$4,368bn	GDP per head in purchasing	
Av. ann. growth in real		power parity (USA=100)	72.7
GDP 1997–2007	1.3%	Economic freedom index	72.5

Origins of GDP[a]		**Components of GDP**	
	% of total		*% of total*
Agriculture	1.5	Private consumption	57.1
Industry, of which:	26.9	Public consumption	17.7
manufacturing	...	Investment	23.5
Services	71.6	Exports	16.1
		Imports	-14.8

Structure of employment

	% of total		*% of labour force*
Agriculture	4	Unemployed 2006	4.1
Industry	28	Av. ann. rate 1995–2006	4.4
Services	68		

Energy

	m TOE		
Total output	99.8	Net energy imports as %	
Total consumption	530.5	of energy use	81
Consumption per head,			
kg oil equivalent	4,152		

Inflation and finance

		av. ann. increase 2001–06	
Consumer price			
inflation 2007	0.1%	Narrow money (M1)	7.1%
Av. ann. inflation 2002–07	0.0%	Broad money	0.4%
Money market rate, 2007	0.47%	Household saving rate, 2007	3.2%

Exchange rates

	end 2007		*December 2007*
¥ per $	114.00	Effective rates	*2000 = 100*
¥ per SDR	180.15	– nominal	82.40
¥ per €	167.82	– real	67.10

Trade

Principal exports		Principal imports	
	$bn fob		*$bn cif*
Transport machinery	156.8	Mineral fuels	160.3
Electrical equipment	138.2	Electrical machinery	138.2
Non-electrical machinery	127.3	General machinery	127.3
Chemicals	58.4	Food	49.0
Metals	30.0	Chemicals	42.1
Total incl. others	**646.4**	Total incl. others	**578.8**

Main export destinations		Main origins of imports	
	% of total		*% of total*
United States	22.5	China	20.5
China	14.3	United States	11.8
South Korea	7.8	Saudi Arabia	6.4
Taiwan	6.8	United Arab Emirates	5.5
Hong Kong	5.6	Australia	4.8

Balance of payments, reserves and aid, $bn

Visible exports fob	615.8	Overall balance	32.0
Visible imports fob	-534.5	Change in reserves	48.4
Trade balance	81.3	Level of reserves	
Invisibles inflows	283.1	end Dec.	895.3
Invisibles outflows	-183.2	No. months of import cover	15.0
Net transfers	-16.7	Official gold holdings, m oz	24.6
Current account balance	170.5	Aid given	11.19
– as % of GDP	3.9	– as % of GDP	0.26
Capital balance	-107.2		

Health and education

Health spending, % of GDP	8.2	Education spending, % of GDP	3.5
Doctors per 1,000 pop.	2.2	Enrolment, %: primary	100
Hospital beds per 1,000 pop.	14.3	secondary	101
Improved-water source access,		tertiary	57
% of pop.	100		

Society

No. of households	48.8m	Colour TVs per 100 households	99.5
Av. no. per household	2.6	Telephone lines per 100 pop.	43.0
Marriages per 1,000 pop.	5.5	Mobile telephone subscribers	
Divorces per 1,000 pop.	2.0	per 100 pop.	77.0
Cost of living, Dec. 2007		Computers per 100 pop.	67.6
New York = 100	125	Internet hosts per 1,000 pop.	287.1

a 2003

KENYA

Area	582,646 sq km	Capital	Nairobi
Arable as % of total land	9	Currency	Kenyan shilling (KSh)

People

Population	35.1m	Life expectancy: men	53.0 yrs
Pop. per sq km	60.2	women	55.2 yrs
Av. ann. growth		Adult literacy	73.6%
in pop. 2010–15	2.55%	Fertility rate (per woman)	4.6
Pop. under 15	42.6%	Urban population	41.6%
Pop. over 60	3.9%		per 1,000 pop.
No. of men per 100 women	99	Crude birth rate	40
Human Development Index	52.1	Crude death rate	11.8

The economy

GDP	KSh1,642bn	GDP per head	$650
GDP	$22.8bn	GDP per head in purchasing	
Av. ann. growth in real		power parity (USA=100)	3.3
GDP 1996–2006	4.2%	Economic freedom index	59.6

Origins of GDP		**Components of GDP**	
	% of total		% of total
Agriculture	24.0	Private consumption	74.3
Industry, of which:	16.7	Public consumption	16.3
manufacturing	10.2	Investment	18.8
Other	54.9	Exports	25.1
		Imports	-37.5

Structure of employment

	% of total		% of labour force
Agriculture	...	Unemployed 2006	...
Industry	...	Av. ann. rate 1995–2006	...
Services	...		

Energy

	m TOE		
Total output	13.9	Net energy imports as %	
Total consumption	17.2	of energy use	19
Consumption per head,			
kg oil equivalent	484		

Inflation and finance

		av. ann. increase 2001–06	
Consumer price			
inflation 2007	9.8%	Narrow money (M1)	18.3%
Av. ann. inflation 2002–07	11.2%	Broad money	13.0%
Treasury bill rate, 2007	6.49%		

Exchange rates

	end 2007		December 2007
KSh per $	62.68	Effective rates	2000 = 100
KSh per SDR	99.04	– nominal	...
KSh per €	92.27	– real	...

Trade

Principal exports		Principal imports	
	$m fob		*$m cif*
Tea	605.0	Industrial supplies	2,231.4
Horticultural products	466.0	Transport equipment	1,327.6
Coffee	126.7	Machinery & other equip.	994.1
Fish products	55.1	Consumer goods	501.2
		Food & drink	443.6
Total incl. others	**3,438.2**	Total incl. others	**7,307.4**

Main export destinations		Main origins of imports	
	% of total		*% of total*
Uganda	15.8	United Arab Emirates	11.9
United Kingdom	10.3	India	8.9
United States	8.2	China	8.4
Netherlands	7.8	Saudi Arabia	8.4

Balance of payments, reserves and debt, $bn

Visible exports fob	3.5	Change in reserves	0.6
Visible imports fob	-6.8	Level of reserves	
Trade balance	-3.3	end Dec.	2.4
Invisibles inflows	2.6	No. months of import cover	3.5
Invisibles outflows	-1.6	Official gold holdings, m oz	0.0
Net transfers	1.8	Foreign debt	6.5
Current account balance	-0.5	– as % of GDP	26
– as % of GDP	-2.3	– as % of total exports	87
Capital balance	0.8	Debt service ratio	7
Overall balance	0.6		

Health and education

Health spending, % of GDP	4.5	Education spending, % of GDP	6.9
Doctors per 1,000 pop.	0.1	Enrolment, %: primary	106
Hospital beds per 1,000 pop.	1.9	secondary	50
Improved-water source access,		tertiary	3
% of pop.	61		

Society

No. of households	7.9m	Colour TVs per 100 households	11.9
Av. no. per household	4.2	Telephone lines per 100 pop.	0.8
Marriages per 1,000 pop.	...	Mobile telephone subscribers	
Divorces per 1,000 pop.	...	per 100 pop.	20.9
Cost of living, Dec. 2007		Computers per 100 pop.	1.4
New York = 100	74	Internet hosts per 1,000 pop.	0.7

LATVIA

Area	63,700 sq km	Capital	Riga
Arable as % of total land	18	Currency	Lats (LVL)

People

Population	2.3m	Life expectancy: men	67.3 yrs
Pop. per sq km	36.1	women	77.7 yrs
Av. ann. growth		Adult literacy	99.7%
in pop. 2010–15	-0.49%	Fertility rate (per woman)	1.3
Pop. under 15	14.4%	Urban population	65.9%
Pop. over 60	22.4%		per 1,000 pop.
No. of men per 100 women	85	Crude birth rate	10
Human Development Index	85.5	Crude death rate	13.6

The economy

GDP	LVL11.3bn	GDP per head	$8,750
GDP	$20.1bn	GDP per head in purchasing	
Av. ann. growth in real		power parity (USA=100)	34.9
GDP 1996–2006	8.8%	Economic freedom index	68.3

Origins of GDP		Components of GDP	
	% of total		% of total
Agriculture	3.5	Private consumption	65.2
Industry, of which:	21.9	Public consumption	16.6
manufacturing	11.8	Investment	32.6
Services	74.6	Exports	44.9
		Imports	-66.3

Structure of employment

	% of total		% of labour force
Agriculture	12	Unemployed 2006	6.8
Industry	26	Av. ann. rate 1996–2006	12.7
Services	62		

Energy

			m TOE
Total output	2.3	Net energy imports as %	
Total consumption	4.7	of energy use	51
Consumption per head,			
kg oil equivalent	2,050		

Inflation and finance

Consumer price		av. ann. increase 2001–06	
inflation 2007	10.1%	Narrow money (M1)	36.3%
Av. ann. inflation 2002–07	6.5%	Broad money	28.9%
Money market rate, 2007	5.07%		

Exchange rates

	end 2007		December 2007
LVL per $	0.48	Effective rates	2000 = 100
LVL per SDR	0.77	– nominal	...
LVL per €	0.71	– real	...

Trade

Principal exports		Principal imports	
	$bn fob		*$bn cif*
Wood & wood products	1.3	Machinery & equipment	2.2
Metals	0.9	Transport equipment	1.6
Machinery & equipment	0.6	Mineral products	1.5
Textiles	0.5	Base metals	1.1
Total incl. others	**5.8**	Total incl. others	**11.1**

Main export destinations		Main origins of imports	
	% of total		*% of total*
Lithuania	14.7	Germany	15.2
Estonia	12.7	Lithuania	13.0
Germany	10.2	Russia	8.0
Russia	9.0	Estonia	7.7
United Kingdom	8.0	Finland	5.6
EU25	75.2	EU25	76.2

Balance of payments, reserves and debt, $bn

Visible exports fob	6.1	Change in reserves	2.2
Visible imports fob	-11.3	Level of reserves	
Trade balance	-5.1	end Dec.	4.5
Invisibles inflows	3.7	No. months of import cover	3.6
Invisibles outflows	-3.6	Official gold holdings, m oz	0.2
Net transfers	0.5	Foreign debt	22.8
Current account balance	-4.5	– as % of GDP	135
– as % of GDP	-22.5	– as % of total exports	266
Capital balance	6.4	Debt service ratio	33
Overall balance	2.0		

Health and education

Health spending, % of GDP	6.4	Education spending, % of GDP	5.1
Doctors per 1,000 pop.	3.1	Enrolment, %: primary	95
Hospital beds per 1,000 pop.	7.7	secondary	99
Improved-water source access,		tertiary	74
% of pop.	99		

Society

No. of households	0.8m	Colour TVs per 100 households	94.8
Av. no. per household	2.9	Telephone lines per 100 pop.	28.6
Marriages per 1,000 pop.	4.7	Mobile telephone subscribers	
Divorces per 1,000 pop.	2.3	per 100 pop.	95.1
Cost of living, Dec. 2007		Computers per 100 pop.	24.6
New York = 100	...	Internet hosts per 1,000 pop.	83.5

LITHUANIA

Area	65,200 sq km	Capital	Vilnius
Arable as % of total land	30	Currency	Litas (LTL)

People

Population	3.4m	Life expectancy: men	67.5 yrs
Pop. per sq km	52.1	women	78.3 yrs
Av. ann. growth		Adult literacy	99.6%
in pop. 2010–15	-0.44%	Fertility rate (per woman)	1.3
Pop. under 15	16.8%	Urban population	66.6%
Pop. over 60	20.3%		per 1,000 pop.
No. of men per 100 women	87	Crude birth rate	9
Human Development Index	86.2	Crude death rate	12.3

The economy

GDP	LTL81.9bn	GDP per head	$8,750
GDP	$29.8bn	GDP per head in purchasing	
Av. ann. growth in real		power parity (USA=100)	35.8
GDP 1996–2006	7.3%	Economic freedom index	70.8

Origins of GDP		Components of GDP	
	% of total		% of total
Agriculture	5.3	Private consumption	65.6
Industry, of which:	33.3	Public consumption	16.9
manufacturing	...	Investment	26.6
Services	61.4	Exports	55.4
		Imports	-67.4

Structure of employment

	% of total		% of labour force
Agriculture	14	Unemployed 2006	5.6
Industry	29	Av. ann. rate 1995–2006	13.5
Services	57		

Energy

		m TOE	
Total output	3.9	Net energy imports as %	
Total consumption	8.6	of energy use	54
Consumption per head,			
kg oil equivalent	2,515		

Inflation and finance

Consumer price		av. ann. increase 2001–06	
inflation 2007	5.7%	Narrow money (M1)	29.8%
Av. ann. inflation 2002–07	2.4%	Broad money	23.4%
Money market rate, 2007	4.18%		

Exchange rates

	end 2007		December 2007
			2000 = 100
LTL per $	2.36	Effective rates	
LTL per SDR	3.73	– nominal	...
LTL per €	3.47	– real	...

Trade

Principal exports		**Principal imports**	
	$bn fob		*$bn cif*
Mineral products	2.4	Machinery & equipment	4.3
Machinery & equipment	2.2	Mineral products	4.2
Transport equipment	1.8	Transport equipment	4.0
Chemicals	1.4	Chemicals	2.3
Total incl. others	**17.1**	Total incl. others	**24.2**

Main export destinations		**Main origins of imports**	
	% of total		*% of total*
Russia	12.8	Russia	24.4
Latvia	11.1	Germany	14.8
Germany	8.7	Poland	9.6
Estonia	6.5	Latvia	4.8
Poland	6.0	Netherlands	3.7
EU25	63.2	EU25	62.4

Balance of payments, reserves and debt, $bn

Visible exports fob	14.2	Change in reserves	2.0
Visible imports fob	-18.4	Level of reserves	
Trade balance	-4.2	end Dec.	5.8
Invisibles inflows	4.2	No. months of import cover	3.1
Invisibles outflows	-3.9	Official gold holdings, m oz	0.2
Net transfers	0.7	Foreign debt	19.0
Current account balance	-3.2	– as % of GDP	79
– as % of GDP	-10.8	– as % of total exports	121
Capital balance	5.0	Debt service ratio	??
Overall balance	1.5		

Health and education

Health spending, % of GDP	5.9	Education spending, % of GDP	5.0
Doctors per 1,000 pop.	4.0	Enrolment, %: primary	95
Hospital beds per 1,000 pop.	8.1	secondary	99
Improved-water source access,		tertiary	76
% of pop.	...		

Society

No. of households	1.4m	Colour TVs per 100 households	93.4
Av. no. per household	2.5	Telephone lines per 100 pop.	23.2
Marriages per 1,000 pop.	6.0	Mobile telephone subscribers	
Divorces per 1,000 pop.	3.4	per 100 pop.	138.1
Cost of living, Dec. 2007		Computers per 100 pop.	18.0
New York = 100	...	Internet hosts per 1,000 pop.	230.5

MALAYSIA

Area	332,665 sq km	Capital	Kuala Lumpur
Arable as % of total land	5	Currency	Malaysian dollar/ringgit (M$)

People

Population	25.8m	Life expectancy: men	72.0 yrs
Pop. per sq km	77.6	women	76.7 yrs
Av. ann. growth		Adult literacy	91.9%
in pop. 2010–15	1.47%	Fertility rate (per woman)	2.4
Pop. under 15	31.4%	Urban population	65.1%
Pop. over 60	6.7%		per 1,000 pop.
No. of men per 100 women	103	Crude birth rate	23
Human Development Index	81.1	Crude death rate	4.5

The economy

GDP	M$573bn	GDP per head	$5,840
GDP	$151bn	GDP per head in purchasing	
Av. ann. growth in real		power parity (USA=100)	28.5
GDP 1997–2007	4.7%	Economic freedom index	64.5

Origins of GDP		Components of GDP	
	% of total		% of total
Agriculture	8.5	Private consumption	45.0
Industry, of which:	46.9	Public consumption	12.0
manufacturing	29.2	Investment	20.9
Services	44.6	Exports	117.0
		Imports	-96.7

Structure of employment

	% of total		% of labour force
Agriculture	15	Unemployed 2004	3.5
Industry	30	Av. ann. rate 1995–2004	3.2
Services	55		

Energy

	m TOE		
Total output	93.9	Net energy imports as %	
Total consumption	61.3	of energy use	-53
Consumption per head,			
kg oil equivalent	2,389		

Inflation and finance

Consumer price		av. ann. increase 2001–06	
inflation 2007	2.0%	Narrow money (M1)	12.0%
Av. ann. inflation 2002–07	2.2%	Broad money	8.3%
Money market rate, 2007	3.50%		

Exchange rates

	end 2007		December 2007
M$ per $	3.31	Effective rates	2000 = 100
M$ per SDR	5.23	– nominal	101.90
M$ per €	4.87	– real	102.20

Trade

Principal exports	$bn fob
Electronics & electrical mach.	60.4
Chemicals & products	9.0
Palm oil	5.9
Total incl. others	**160.7**

Principal imports	$bn cif
Intermediate goods	69.8
Capital goods & transport equipment	13.6
Consumption goods	5.8
Total incl. others	**131.2**

Main export destinations	% of total
United States	18.8
Singapore	15.4
Japan	8.9
China	7.2
Thailand	5.3

Main origins of imports	% of total
Japan	13.2
United States	12.5
China	12.1
Singapore	11.7
Thailand	5.5

Balance of payments, reserves and debt, $bn

Visible exports fob	160.8	Change in reserves	12.4
Visible imports fob	-124.1	Level of reserves	
Trade balance	36.7	end Dec.	82.9
Invisibles inflows	30.3	No. months of import cover	6.2
Invisibles outflows	-36.9	Official gold holdings, m oz	1.2
Net transfers	-4.6	Foreign debt	52.5
Current account balance	25.5	– as % of GDP	39
– as % of GDP	16.9	– as % of total exports	31
Capital balance	-11.9	Debt service ratio	4
Overall balance	6.9		

Health and education

Health spending, % of GDP	4.2	Education spending, % of GDP	6.2
Doctors per 1,000 pop.	0.8	Enrolment, %: primary	100
Hospital beds per 1,000 pop.	1.8	secondary	69
Improved-water source access, % of pop.	99	tertiary	29

Society

No. of households	5.7m	Colour TVs per 100 households	94.0
Av. no. per household	4.7	Telephone lines per 100 pop.	16.8
Marriages per 1,000 pop.	5.9	Mobile telephone subscribers	
Divorces per 1,000 pop.	...	per 100 pop.	75.5
Cost of living, Dec. 2007		Computers per 100 pop.	21.8
New York = 100	71	Internet hosts per 1,000 pop.	14.9

MEXICO

Area	1,972,545 sq km	Capital	Mexico city
Arable as % of total land	13	Currency	Mexican peso (PS)

People

Population	108.3m	Life expectancy: men	73.7 yrs
Pop. per sq km	54.9	women	78.6 yrs
Av. ann. growth		Adult literacy	92.4%
in pop. 2010–15	0.97%	Fertility rate (per woman)	2.0
Pop. under 15	30.8%	Urban population	76.0%
Pop. over 60	8.4%		per 1,000 pop.
No. of men per 100 women	95	Crude birth rate	21
Human Development Index	82.9	Crude death rate	4.8

The economy

GDP	PS9,155bn	GDP per head	$7,750
GDP	$839bn	GDP per head in purchasing	
Av. ann. growth in real		power parity (USA=100)	27.7
GDP 1997–2007	3.6%	Economic freedom index	66.4

Origins of GDP		Components of GDP	
	% of total		% of total
Agriculture	3.9	Private consumption	67.6
Industry, of which:	26.7	Public consumption	11.7
manufacturing & mining	18.0	Investment	20.4
Services	69.4	Exports	31.8
		Imports	-33.1

Structure of employment

	% of total		% of labour force
Agriculture	15	Unemployed 2006	3.2
Industry	26	Av. ann. rate 1995–2006	2.8
Services	59		

Energy

	m TOE		
Total output	259.2	Net energy imports as %	
Total consumption	176.5	of energy use	-47
Consumption per head,			
kg oil equivalent	1,712		

Inflation and finance

Consumer price		av. ann. increase 2001–06	
inflation 2007	4.0%	Narrow money (M1)	13.4%
Av. ann. inflation 2002–07	4.2%	Broad money	9.7%
Money market rate, 2007	7.66%		

Exchange rates

	end 2007		December 2007
PS per $	10.87	Effective rates	2000 = 100
PS per SDR	17.17	– nominal	...
PS per €	16.00	– real	...

Trade

Principal exports		Principal imports	
	$bn fob		*$bn fob*
Manufactured products	202.7	Intermediate goods	188.6
(Maquiladora[a]	*111.8)*	*(Maquiladora[a]*	*87.5)*
Crude oil & products	39.0	Consumer goods	36.9
Agricultural products	6.9	Capital goods	30.5
Total incl. others	**249.9**	Total	**256.1**

Main export destinations		Main origins of imports	
	% of total		*% of total*
United States	84.7	United States	50.9
Canada	2.1	China	9.5
Germany	1.3	Japan	5.9
Spain	1.2	South Korea	4.2

Balance of payments, reserves and debt, $bn

Visible exports fob	250.0	Change in reserves	2.2
Visible imports fob	-256.1	Level of reserves	
Trade balance	-6.1	end Dec.	76.3
Invisibles inflows	22.8	No. months of import cover	3.1
Invisibles outflows	-42.8	Official gold holdings, m oz	0.1
Net transfers	24.1	Foreign debt	160.7
Current account balance	-2.0	– as % of GDP	21
– as % of GDP	-0.2	– as % of total exports	62
Capital balance	-2.2	Debt service ratio	19
Overall balance	-1.3		

Health and education

Health spending, % of GDP	6.4	Education spending, % of GDP	5.5
Doctors per 1,000 pop.	1.7	Enrolment, %: primary	113
Hospital beds per 1,000 pop.	1.0	secondary	87
Improved-water source access,		tertiary	26
% of pop.	97		

Society

No. of households	25.1m	Colour TVs per 100 households	91.8
Av. no. per household	4.3	Telephone lines per 100 pop.	18.3
Marriages per 1,000 pop.	5.7	Mobile telephone subscribers	
Divorces per 1,000 pop.	0.7	per 100 pop.	52.0
Cost of living, Dec. 2007		Computers per 100 pop.	13.6
New York = 100	80	Internet hosts per 1,000 pop.	93.0

a Manufacturing assembly plants near the Mexican-US border where goods for processing may be imported duty-free and all output is exported.

MOROCCO

Area	446,550 sq km	Capital	Rabat
Arable as % of total land	19	Currency	Dirham (Dh)

People

Population	31.9m	Life expectancy: men	69.0 yrs
Pop. per sq km	71.4	women	73.4 yrs
Av. ann. growth		Adult literacy	52.3%
in pop. 2010–15	1.17%	Fertility rate (per woman)	2.3
Pop. under 15	30.3%	Urban population	58.8%
Pop. over 60	7.5%		per 1,000 pop.
No. of men per 100 women	97	Crude birth rate	21
Human Development Index	64.6	Crude death rate	5.8

The economy

GDP	Dh575bn	GDP per head	$2,050
GDP	$65.4bn	GDP per head in purchasing	
Av. ann. growth in real		power parity (USA=100)	8.9
GDP 1996–2006	5.0%	Economic freedom index	56.4

Origins of GDP		Components of GDP	
	% of total		% of total
Agriculture	15.2	Private consumption	61.3
Industry, of which:	38.7	Public consumption	20.1
manufacturing	19.7	Investment	24.1
Services	46.1	Exports	30.9
		Imports	-36.9

Structure of employment

	% of total		% of labour force
Agriculture	45	Unemployed 2006	9.7
Industry	20	Av. ann. rate 1995–2006	14.3
Services	35		

Energy

	m TOE		
Total output	1.0	Net energy imports as %	
Total consumption	13.5	of energy use	93
Consumption per head,			
kg oil equivalent	458		

Inflation and finance

Consumer price		av. ann. increase 2001–06	
inflation 2007	2.0%	Narrow money (M1)	11.8%
Av. ann. inflation 2002–07	1.8%	Broad money	10.7%
Money market rate, 2007	3.31%		

Exchange rates

	end 2007		December 2007
		Effective rates	2000 = 100
Dh per $	7.71	Effective rates	2000 = 100
Dh per SDR	12.19	– nominal	95.60
Dh per €	11.35	– real	92.20

Trade

Principal exports		Principal imports	
	$bn fob		*$bn cif*
Textiles	2.4	Semi-finished goods	5.6
Phosphoric acid	0.9	Capital goods	5.3
Electrical components	0.7	Energy & lubricants	5.1
Phosphate rock	0.6	Consumer goods	4.7
Citrus fruits	0.3	Food, drink & tobacco	1.8
Total incl. others	**11.5**	Total incl. others	**22.5**

Main export destinations		Main origins of imports	
	% of total		*% of total*
France	21.4	France	17.4
Spain	20.5	Spain	13.4
United Kingdom	4.9	Saudi Arabia	6.9
Italy	4.7	China	6.8

Balance of payments, reserves and debt, $bn

Visible exports fob	11.9	Change in reserves	4.2
Visible imports fob	-21.3	Level of reserves	
Trade balance	-9.4	end Dec.	20.8
Invisibles inflows	10.6	No. months of import cover	9.2
Invisibles outflows	-5.6	Official gold holdings, m oz	0.7
Net transfers	6.3	Foreign debt	18.5
Current account balance	1.8	– as % of GDP	30
– as % of GDP	2.7	– as % of total exports	72
Capital balance	-0.6	Debt service ratio	12
Overall balance	0.7		

Health and education

Health spending, % of GDP	5.3	Education spending, % of GDP	6.8
Doctors per 1,000 pop.	0.6	Enrolment, %: primary	106
Hospital beds per 1,000 pop.	0.9	secondary	52
Improved-water source access,		tertiary	12
% of pop.	81		

Society

No. of households	6.1m	Colour TVs per 100 households	73.0
Av. no. per household	5.2	Telephone lines per 100 pop.	4.1
Marriages per 1,000 pop.	...	Mobile telephone subscribers	
Divorces per 1,000 pop.	...	per 100 pop.	52.1
Cost of living, Dec. 2007		Computers per 100 pop.	2.5
New York = 100	77	Internet hosts per 1,000 pop.	8.6

NETHERLANDS

Area[a]	41,526 sq km	Capital	Amsterdam
Arable as % of total land	27	Currency	Euro (€)

People

Population	16.4m	Life expectancy: men	77.5 yrs
Pop. per sq km	394.9	women	81.9 yrs
Av. ann. growth		Adult literacy	...
in pop. 2010–15	0.15%	Fertility rate (per woman)	1.7
Pop. under 15	18.4%	Urban population	66.8%
Pop. over 60	19.3%		per 1,000 pop.
No. of men per 100 women	98	Crude birth rate	11
Human Development Index	95.3	Crude death rate	8.6

The economy

GDP	€528bn	GDP per head	$40,380
GDP	$662bn	GDP per head in purchasing	
Av. ann. growth in real		power parity (USA=100)	83.2
GDP 1996–2006	2.8%	Economic freedom index	76.8

Origins of GDP		Components of GDP	
	% of total		% of total
Agriculture	2.2	Private consumption	47.4
Industry, of which:	24.1	Public consumption	25.3
manufacturing	...	Investment	19.9
Services	73.7	Exports	75.3
		Imports	-67.3

Structure of employment

	% of total		% of labour force
Agriculture	3	Unemployed 2005	5.2
Industry	20	Av. ann. rate 1995–2005	4.6
Services	77		

Energy

	m TOE		
Total output	61.9	Net energy imports as %	
Total consumption	81.8	of energy use	24
Consumption per head,			
kg oil equivalent	5,015		

Inflation and finance

Consumer price		av. ann. increase 2001–06	
inflation 2007	1.6%	Euro area:	
Av. ann. inflation 2002–07	1.5%	Narrow money (M1)	10.5%
Deposit rate, households, 2007	3.86%	Broad money	7.4%
		Household saving rate, 2007	6.1%

Exchange rates

	end 2007		December 2007
€ per $	0.68	Effective rates	2000 = 100
€ per SDR	1.07	– nominal	114.20
		– real	114.40

Trade

Principal exports		Principal imports	
	$bn fob		*$bn cif*
Machinery & transport		Machinery & transport	
equipment	123.0	equipment	122.6
Food & live animals	43.4	Food & live animals	26.2
Total incl. others	**400.8**	Total incl. others	**358.6**

Main export destinations		Main origins of imports	
	% of total		*% of total*
Germany	23.8	Germany	19.6
Belgium	11.9	Belgium	11.1
United Kingdom	10.0	United States	8.0
France	9.7	China	6.5
EU25	76.9	EU25	55.3

Balance of payments, reserves and aid, $bn

Visible exports fob	386.9	Overall balance	0.8
Visible imports fob	-341.7	Change in reserves	3.5
Trade balance	45.2	Level of reserves	
Invisibles inflows	213.4	end Dec.	23.9
Invisibles outflows	-190.3	No. months of import cover	0.5
Net transfers	-12.5	Official gold holdings, m oz	20.6
Current account balance	55.8	Aid given	5.45
– as % of GDP	8.4	– as % of GDP	0.82
Capital balance	-56.6		

Health and education

Health spending, % of GDP	9.2	Education spending, % of GDP	5.3
Doctors per 1,000 pop.	3.7	Enrolment, %: primary	107
Hospital beds per 1,000 pop.	5.0	secondary	118
Improved-water source access,		tertiary	60
% of pop.	100		

Society

No. of households	7.2m	Colour TVs per 100 households	98.3
Av. no. per household	2.3	Telephone lines per 100 pop.	46.6
Marriages per 1,000 pop.	4.3	Mobile telephone subscribers	
Divorces per 1,000 pop.	1.8	per 100 pop.	106.0
Cost of living, Dec. 2007		Computers per 100 pop.	85.4
New York = 100	108	Internet hosts per 1,000 pop.	642.7

a Includes water.

NEW ZEALAND

Area	270,534 sq km	Capital	Wellington
Arable as % of total land	6	Currency	New Zealand dollar (NZ$)

People

Population	4.1m	Life expectancy: men		78.2 yrs
Pop. per sq km	15.2		women	82.2 yrs
Av. ann. growth		Adult literacy		...
in pop. 2010–15	0.79%	Fertility rate (per woman)		1.9
Pop. under 15	21.5%	Urban population		86.0%
Pop. over 60	16.6%			per 1,000 pop.
No. of men per 100 women	97	Crude birth rate		14
Human Development Index	94.3	Crude death rate		7.1

The economy

GDP	NZ$161bn	GDP per head	$25,490
GDP	$105bn	GDP per head in purchasing	
Av. ann. growth in real		power parity (USA=100)	58.0
GDP 1996–2006	3.5%	Economic freedom index	80.2

Origins of GDP

Components of GDP

	% of total		% of total
Agriculture & mining	4.6	Private consumption	59.9
Industry	26.9	Public consumption	18.6
Services	68.5	Investment	23.2
		Exports	29.2
		Imports	-30.8

Structure of employment

	% of total		% of labour force
Agriculture	7	Unemployed 2006	3.8
Industry	22	Av. ann. rate 1995–2006	5.5
Services	71		

Energy

	m TOE		
Total output	12.2	Net energy imports as %	
Total consumption	16.9	of energy use	28
Consumption per head,			
kg oil equivalent	4,090		

Inflation and finance

Consumer price		av. ann. increase 2001–06	
inflation 2007	2.4%	Narrow money (M1)	11.0%
Av. ann. inflation 2002–07	2.6%	Broad money	9.8%
Money market rate, 2007	7.93%		

Exchange rates

	end 2007		December 2007
NZ$ per $	1.29	Effective rates	2000 = 100
NZ$ per SDR	2.04	– nominal	134.10
NZ$ per €	1.90	– real	138.80

Trade

Principal exports

	$bn fob
Dairy produce	4.0
Meat	3.0
Forestry products	1.4
Wool	0.4
Total incl. others	**22.4**

Principal imports

	$bn cif
Machinery & electrical equipment	5.3
Transport equipment	3.9
Mineral fuels	3.2
Total incl. others	**26.4**

Main export destinations

	% of total
Australia	20.6
United States	13.2
Japan	10.4
China	5.5

Main origins of imports

	% of total
Australia	20.1
China	12.2
United States	12.1
Japan	9.1

Balance of payments, reserves and aid, $bn

Visible exports fob	22.5	Overall balance	4.3
Visible imports fob	-24.6	Change in reserves	5.2
Trade balance	-2.1	Level of reserves	
Invisibles inflows	9.3	end Dec.	14.1
Invisibles outflows	-17.1	No. months of import cover	4.0
Net transfers	0.5	Official gold holdings, m oz	0.0
Current account balance	-9.4	Aid given	0.26
– as % of GDP	-9.0	– as % of GDP	0.25
Capital balance	13.4		

Health and education

Health spending, % of GDP	9.2	Education spending, % of GDP	6.2
Doctors per 1,000 pop.	2.1	Enrolment, %: primary	102
Hospital beds per 1,000 pop.	6.0	secondary	120
Improved-water source access, % of pop.	...	tertiary	80

Society

No. of households	1.5m	Colour TVs per 100 households	98.0
Av. no. per household	2.6	Telephone lines per 100 pop.	44.1
Marriages per 1,000 pop.	5.1	Mobile telephone subscribers	
Divorces per 1,000 pop.	2.7	per 100 pop.	94.0
Cost of living, Dec. 2007		Computers per 100 pop.	50.2
New York = 100	93	Internet hosts per 1,000 pop.	411.6

NIGERIA

Area	923,768 sq km	Capital	Abuja
Arable as % of total land	35	Currency	Naira (N)

People

Population	134.4m	Life expectancy: men	46.4 yrs
Pop. per sq km	145.5	women	47.3 yrs
Av. ann. growth		Adult literacy	72.0%
in pop. 2010–15	2.09%	Fertility rate (per woman)	4.7
Pop. under 15	44.3%	Urban population	48.3%
Pop. over 60	4.6%		per 1,000 pop.
No. of men per 100 women	100	Crude birth rate	43
Human Development Index	47.0	Crude death rate	16.8

The economy

GDP	N14,693bn	GDP per head	$860
GDP	$115bn	GDP per head in purchasing	
Av. ann. growth in real		power parity (USA=100)	3.7
GDP 1996–2006	8.4%	Economic freedom index	55.5

Origins of GDP		Components of GDP	
	% of total		% of total
Agriculture	23	Private consumption	40
Industry, of which:	57	Public consumption	21
manufacturing	4	Investment	21
Services	20	Exports	53
		Imports	-35

Structure of employment

	% of total		% of labour force
Agriculture	...	Unemployed 2001	3.9
Industry	...	Av. ann. rate 1995–2001	3.7
Services	...		

Energy

	m TOE		
Total output	231.8	Net energy imports as %	
Total consumption	103.8	of energy use	-123
Consumption per head,			
kg oil equivalent	734		

Inflation and finance

		av. ann. increase 2001–05	
Consumer price			
inflation 2007	5.4%	Narrow money (M1)	17.2%
Av. ann. inflation 2002–07	12.0%	Broad money	18.9%
Treasury bill rate 2006	9.99%		

Exchange rates

	end 2007		December 2007
N per $	117.97	Effective rates	2000 = 100
N per SDR	186.42	– nominal	63.70
N per €	173.66	– real	128.20

Trade

Principal exports[a]		Principal imports[a]	
	$bn fob		*$bn cif*
Oil	45.4	Manufactured goods	6.1
Gas	4.6	Machinery & transport equipment	4.6
		Chemicals	4.2
		Agric products & foodstuffs	1.3
Total incl. others	**50.2**	Total incl. others	**25.6**

Main export destinations		Main origins of imports	
	% of total		*% of total*
United States	58.4	China	10.6
Spain	9.5	United States	8.3
Brazil	7.4	Netherlands	5.9
France	5.0	United Kingdom	5.7

Balance of payments, reserves and debt, $bn

Visible exports fob	59.1	Change in reserves	14.1
Visible imports fob	-31.1	Level of reserves	
Trade balance	28.0	end Dec.	42.7
Net invisibles	-17.7	No. months of import cover	9.6
Net transfers	3.4	Official gold holdings, m oz	0.7
Current account balance	13.8	Foreign debt	7.7
– as % of GDP	12.0	– as % of GDP	9
Capital balance	1.8	– as % of total exports	13
Overall balance	13.9	Debt service ratio[a]	16

Health and education

Health spending, % of GDP	3.9	Education spending, % of GDP	...
Doctors per 1,000 pop.	0.3	Enrolment, %: primary	96
Hospital beds per 1,000 pop.	...	secondary	32
Improved-water source access, % of pop.	48	tertiary	10

Society

No. of households	27.5m	Colour TVs per 100 households	32.0
Av. no. per household	5.3	Telephone lines per 100 pop.	1.3
Marriages per 1,000 pop.	...	Mobile telephone subscribers	
Divorces per 1,000 pop.	...	per 100 pop.	24.1
Cost of living, Dec. 2007		Computers per 100 pop.	0.8
New York = 100	84	Internet hosts per 1,000 pop.	...

a 2005

NORWAY

Area	323,878 sq km	Capital	Oslo
Arable as % of total land	3	Currency	Norwegian krone (Nkr)

People

Population	4.6m	Life expectancy: men		77.8 yrs
Pop. per sq km	14.2	women		82.5 yrs
Av. ann. growth		Adult literacy		...
in pop. 2010–15	0.60%	Fertility rate (per woman)		1.9
Pop. under 15	19.6%	Urban population		80.5%
Pop. over 60	19.7%			per 1,000 pop.
No. of men per 100 women	99	Crude birth rate		13
Human Development Index	96.8	Crude death rate		9.1

The economy

GDP	Nkr2,148bn	GDP per head	$72,810
GDP	$335bn	GDP per head in purchasing	
Av. ann. growth in real		power parity (USA=100)	113.9
GDP 1997–2007	2.8%	Economic freedom index	69.0

Origins of GDP		Components of GDP	
	% of total		% of total
Agriculture	2.4	Private consumption	40.9
Industry, of which:	43.5	Public consumption	19.2
manufacturing	17.9	Investment	18.9
Services	54.1	Exports	46.5
		Imports	-28.3

Structure of employment

	% of total		% of labour force
Agriculture	3	Unemployed 2006	3.4
Industry	21	Av. ann. rate 1995–2006	4.0
Services	76		

Energy

	m TOE		
Total output	233.7	Net energy imports as %	
Total consumption	32.1	of energy use	-627
Consumption per head,			
kg oil equivalent	6,948		

Inflation and finance

Consumer price		av. ann. increase 2001–06	
inflation 2007	0.7%	Narrow money (M1)	12.1%
Av. ann. inflation 2002–07	1.5%	Broad money	8.5%
Interbank rate, 2007	4.95%	Household saving rate, 2007	0.8%

Exchange rates

	end 2007		December 2007
Nkr per $	5.41	Effective rates	2000 = 100
Nkr per SDR	8.55	– nominal	116.50
Nkr per €	7.96	– real	115.50

Trade

Principal exports		Principal imports	
	$bn fob		*$bn cif*
Oil, gas & products	82.9	Machinery & transport equip.	24.5
Metals	9.7	Metals	4.1
Machinery & transport equip.	8.0	Chemicals	3.6
Food, drink & tobacco	5.9	Refined petroleum products	3.0
Total incl. others	**122.1**	Total incl. others	**64.3**

Main export destinations		Main origins of imports	
	% of total		*% of total*
United Kingdom	26.6	Sweden	15.0
Germany	12.2	Germany	13.5
Netherlands	10.4	Denmark	6.9
France	8.2	United Kingdom	6.4
Sweden	6.4	China	5.7
United States	5.9	United States	5.3
EU25	81.8	EU25	67.5

Balance of payments, reserves and aid, $bn

Visible exports fob	122.8	Overall balance	5.5
Visible imports fob	-62.9	Change in reserves	9.9
Trade balance	59.9	Level of reserves	
Invisibles inflows	64.3	end Dec.	56.8
Invisibles outflows	-63.6	No. months of import cover	5.4
Net transfers	-2.3	Official gold holdings, m oz	0.0
Current account balance	58.3	Aid given	2.95
– as % of GDP	17.4	– as % of GDP	0.88
Capital balance	-40.2		

Health and education

Health spending, % of GDP	9.0	Education spending, % of GDP	7.2
Doctors per 1,000 pop.	3.8	Enrolment, %: primary	98
Hospital beds per 1,000 pop.	4.2	secondary	113
Improved-water source access,		tertiary	78
% of pop.	100		

Society

No. of households	2.0m	Colour TVs per 100 households	98.2
Av. no. per household	2.3	Telephone lines per 100 pop.	44.3
Marriages per 1,000 pop.	4.8	Mobile telephone subscribers	
Divorces per 1,000 pop.	2.5	per 100 pop.	108.6
Cost of living, Dec. 2007		Computers per 100 pop.	59.4
New York = 100	148	Internet hosts per 1,000 pop.	592.4

PAKISTAN

Area	803,940 sq km	Capital	Islamabad
Arable as % of total land	28	Currency	Pakistan rupee (PRs)

People

Population	161.2m	Life expectancy:	men	65.2 yrs
Pop. per sq km	200.5		women	65.8 yrs
Av. ann. growth		Adult literacy		54.9%
in pop. 2010–15	1.90%	Fertility rate (per woman)		3.2
Pop. under 15	37.2%	Urban population		34.8%
Pop. over 60	5.9%			*per 1,000 pop.*
No. of men per 100 women	106	Crude birth rate		31
Human Development Index	55.1	Crude death rate		7.1

The economy

GDP	PRs7,594bn	GDP per head	$790
GDP	$127bn	GDP per head in purchasing	
Av. ann. growth in real		power parity (USA=100)	5.4
GDP 1996–2006	5.4%	Economic freedom index	56.8

Origins of GDP[a]		Components of GDP[a]	
	% of total		*% of total*
Agriculture	19.6	Private consumption	74.9
Industry, of which:	26.8	Public consumption	11.2
manufacturing	17.7	Investment	21.4
Services	53.7	Exports	13.9
		Imports	-22.2

Structure of employment

	% of total		*% of labour force*
Agriculture	43	Unemployed 2006	6.2
Industry	20	Av. ann. rate 1995–2006	6.5
Services	37		

Energy

	m TOE		
Total output	61.3	Net energy imports as %	
Total consumption	76.3	of energy use	20
Consumption per head,			
kg oil equivalent	490		

Inflation and finance

		av. ann. increase 2001–06	
Consumer price			
inflation 2007	7.6%	Narrow money (M1)	27.9%
Av. ann. inflation 2002–07	7.0%	Broad money	17.2%
Money market rate, 2007	9.30%		

Exchange rates

	end 2007		*December 2007*
PRs per $	61.22	Effective rates	*2000 = 100*
PRs per SDR	96.74	– nominal	72.88
PRs per €	90.12	– real	93.62

Trade[b]

Principal exports		Principal imports	
	$bn fob		*$bn fob*
Cotton fabrics	2.1	Machinery & transport equip.	8.3
Cotton yarn & thread	1.4	Fuels etc	6.9
Rice	1.2	Chemicals	4.2
Raw cotton	0.1	Palm oil	0.8
Total incl. others	**16.9**	Total incl. others	**29.8**

Main export destinations		Main origins of imports	
	% of total		*% of total*
United States	21.3	China	15.6
United Arab Emirates	9.2	Saudi Arabia	11.9
Afghanistan	7.8	United Arab Emirates	11.0
China	5.4	United States	7.3
United Kingdom	5.2	Japan	6.5

Balance of payments, reserves and debt, $bn

Visible exports fob	17.0	Change in reserves	1.8
Visible imports fob	-26.7	Level of reserves	
Trade balance	-9.7	end Dec.	12.9
Invisibles inflows	4.4	No. months of import cover	4.0
Invisibles outflows	-12.4	Official gold holdings, m oz	2.1
Net transfers	10.9	Foreign debt	35.9
Current account balance	-6.8	– as % of GDP	26
– as % of GDP	-5.4	– as % of total exports	123
Capital balance	7.6	Debt service ratio	9
Overall balance	1.6		

Health and education

Health spending, % of GDP	2.1	Education spending, % of GDP	2.6
Doctors per 1,000 pop.	0.8	Enrolment, %: primary	84
Hospital beds per 1,000 pop.	0.7	secondary	30
Improved-water source access,		tertiary	5
% of pop.	91		

Society

No. of households	22.7m	Colour TVs per 100 households	61.8
Av. no. per household	7.1	Telephone lines per 100 pop.	3.3
Marriages per 1,000 pop.	...	Mobile telephone subscribers	
Divorces per 1,000 pop.	...	per 100 pop.	22.0
Cost of living, Dec. 2007		Computers per 100 pop.	0.5
New York = 100	44	Internet hosts per 1,000 pop.	1.1

a Fiscal year ending June 30, 2007.
b Fiscal year ending June 30, 2006.

PERU

Area	1,285,216 sq km	Capital	Lima
Arable as % of total land	3	Currency	Nuevo Sol (New Sol)

People

Population	28.4m	Life expectancy: men	68.9 yrs
Pop. per sq km	22.1	women	74.0 yrs
Av. ann. growth		Adult literacy	90.5%
in pop. 2010–15	1.26%	Fertility rate (per woman)	2.4
Pop. under 15	31.8%	Urban population	74.6%
Pop. over 60	8.1%		*per 1,000 pop.*
No. of men per 100 women	100	Crude birth rate	21
Human Development Index	77.3	Crude death rate	6.1

The economy

GDP	New Soles 303bn	GDP per head	$3,250
GDP	$92.4bn	GDP per head in purchasing	
Av. ann. growth in real		power parity (USA=100)	16.1
GDP 1997–2007	4.5%	Economic freedom index	63.5

Origins of GDP

Components of GDP

	% of total		*% of total*
Agriculture	8.7	Private consumption	61.5
Industry, of which:	26.2	Public consumption	9.6
manufacturing	15.1	Investment	18.5
Services	53.2	Exports	28.7
		Imports	-19.6

Structure of employment

	% of total		*% of labour force*
Agriculture	1	Unemployed 2006	8.8
Industry	23	Av. ann. rate 1996–2006	8.8
Services	76		

Energy

	m TOE		
Total output	10.8	Net energy imports as %	
Total consumption	13.8	of energy use	22
Consumption per head,			
kg oil equivalent	506		

Inflation and finance

Consumer price		*av. ann. increase 2001–06*	
inflation 2007	1.8%	Narrow money (M1)	8.8%
Av. ann. inflation 2002–07	2.3%	Broad money	6.7%
Money market rate, 2007	4.99%		

Exchange rates

	end 2007		*December 2007*
New Soles per $	3.00	Effective rates	*2000 = 100*
New Soles per SDR	4.73	– nominal	...
New Soles per €	4.42	– real	...

Trade

Principal exports		Principal imports	
	$bn fob		*$bn fob*
Copper	6.1	Intermediate goods	8.0
Gold	4.0	Capital goods	4.1
Zinc	2.0	Consumer goods	3.2
Fishmeal	1.3	Other goods	0.1
Total incl. others	**23.8**	Total incl. others	**14.9**

Main export destinations		Main origins of imports	
	% of total		*% of total*
United States	23.6	United States	21.6
China	11.0	Brazil	8.7
Chile	7.7	Ecuador	8.1
Canada	5.5	China	7.5

Balance of payments, reserves and debt, $bn

Visible exports fob	23.8	Change in reserves	3.3
Visible imports fob	-14.9	Level of reserves	
Trade balance	8.9	end Dec.	17.4
Invisibles inflows	3.5	No. months of import cover	7.8
Invisibles outflows	-12.0	Official gold holdings, m oz	1.1
Net transfers	2.2	Foreign debt	28.2
Current account balance	2.6	– as % of GDP	42
– as % of GDP	2.8	– as % of total exports	140
Capital balance	1.1	Debt service ratio	13
Overall balance	3.2		

Health and education

Health spending, % of GDP	4.3	Education spending, % of GDP	2.5
Doctors per 1,000 pop.	1.2	Enrolment, %: primary	116
Hospital beds per 1,000 pop.	1.1	secondary	94
Improved-water source access,		tertiary	35
% of pop.	83		

Society

No. of households	5.9m	Colour TVs per 100 households	55.7
Av. no. per household	4.8	Telephone lines per 100 pop.	8.5
Marriages per 1,000 pop.	2.8	Mobile telephone subscribers	
Divorces per 1,000 pop.	...	per 100 pop.	30.9
Cost of living, Dec. 2007		Computers per 100 pop.	10.3
New York = 100	61	Internet hosts per 1,000 pop.	9.6

PHILIPPINES

Area	300,000 sq km	Capital	Manila
Arable as % of total land	19	Currency	Philippine peso (P)

People

Population	84.5m	Life expectancy:	men	69.5 yrs
Pop. per sq km	281.7		women	73.9 yrs
Av. ann. growth		Adult literacy		93.4%
in pop. 2010–15	1.67%	Fertility rate (per woman)		2.9
Pop. under 15	36.2%	Urban population		62.6%
Pop. over 60	6.0%			per 1,000 pop.
No. of men per 100 women	101	Crude birth rate		27
Human Development Index	77.1	Crude death rate		4.8

The economy

GDP	P6,033bn	GDP per head	$1,390
GDP	$118bn	GDP per head in purchasing	
Av. ann. growth in real		power parity (USA=100)	7.2
GDP 1997–2007	4.9%	Economic freedom index	56.9

Origins of GDP		Components of GDP	
	% of total		% of total
Agriculture	14.2	Private consumption	70.1
Industry, of which:	31.6	Public consumption	9.7
manufacturing	22.9	Investment	13.8
Services	54.2	Exports	46.4
		Imports	-47.6

Structure of employment

	% of total		% of labour force
Agriculture	37	Unemployed 2006	7.3
Industry	15	Av. ann. rate 1995–2006	9.1
Services	48		

Energy

	m TOE		
Total output	24.2	Net energy imports as %	
Total consumption	44.7	of energy use	46
Consumption per head,			
kg oil equivalent	528		

Inflation and finance

Consumer price		av. ann. increase 2001–06	
inflation 2007	2.8%	Narrow money (M1)	14.5%
Av. ann. inflation 2002–07	5.2%	Broad money	9.9%
Money market rate, 2007	7.02%		

Exchange rates

	end 2007		December 2007
P per $	41.40	Effective rates	2000 = 100
P per SDR	65.42	– nominal	96.20
P per €	60.94	– real	121.80

Trade

Principal exports		Principal imports	
	$bn fob		*$bn fob*
Electrical & electronic		Electronic products	24.4
equipment	29.6	Fuels	8.2
Clothing	2.6	Industrial machinery	2.0
Mineral products	2.1	Transport equipment	2.0
Agricultural products	1.6	Textile fabrics	1.1
Total incl. others	**47.1**	Total incl. others	**53.8**

Main export destinations		Main origins of imports	
	% of total		*% of total*
United States	18.3	United States	17.9
Japan	16.5	Japan	14.7
Netherlands	10.1	Singapore	8.7
China	9.8	China	7.5
Hong Kong	7.9	Taiwan	7.4

Balance of payments, reserves and debt, $bn

Visible exports fob	46.5	Change in reserves	4.5
Visible imports fob	-53.3	Level of reserves	
Trade balance	-6.8	end Dec.	23.0
Invisibles inflows	10.8	No. months of import cover	4.3
Invisibles outflows	-11.3	Official gold holdings, m oz	4.6
Net transfers	13.2	Foreign debt	60.3
Current account balance	5.9	– as % of GDP	57
– as % of GDP	5.0	– as % of total exports	101
Capital balance	-0.5	Debt service ratio	20
Overall balance	4.7		

Health and education

Health spending, % of GDP	3.2	Education spending, % of GDP	2.5
Doctors per 1,000 pop.	1.1	Enrolment, %: primary	110
Hospital beds per 1,000 pop.	1.2	secondary	83
Improved-water source access,		tertiary	28
% of pop.	85		

Society

No. of households	17.5m	Colour TVs per 100 households	80.0
Av. no. per household	4.9	Telephone lines per 100 pop.	4.3
Marriages per 1,000 pop.	6.7	Mobile telephone subscribers	
Divorces per 1,000 pop.	...	per 100 pop.	50.8
Cost of living, Dec. 2007		Computers per 100 pop.	5.3
New York = 100	46	Internet hosts per 1,000 pop.	3.2

POLAND

Area	312,683 sq km	Capital	Warsaw
Arable as % of total land	40	Currency	Zloty (Zl)

People

Population	38.5m	Life expectancy:	men	71.3 yrs
Pop. per sq km	123.1		women	79.8 yrs
Av. ann. growth		Adult literacy		...
in pop. 2010–15	-0.17%	Fertility rate (per woman)		1.3
Pop. under 15	16.3%	Urban population		62.0%
Pop. over 60	17.2%			per 1,000 pop.
No. of men per 100 women	93	Crude birth rate		10
Human Development Index	87.0	Crude death rate		10.0

The economy

GDP	Zl1,051bn	GDP per head	$8,800
GDP	$339bn	GDP per head in purchasing	
Av. ann. growth in real		power parity (USA=100)	33.7
GDP 1997–2007	4.7%	Economic freedom index	59.5

Origins of GDP

	% of total
Agriculture	4.3
Industry, of which:	31.2
manufacturing	18.9
Services	64.5

Components of GDP

	% of total
Private consumption	61.1
Public consumption	18.3
Investment	19.7
Exports	40.3
Imports	-41.7

Structure of employment

	% of total		% of labour force
Agriculture	17	Unemployed 2006	13.8
Industry	29	Av. ann. rate 1995–2006	15.5
Services	54		

Energy

	m TOE		
Total output	78.6	Net energy imports as %	
Total consumption	93	of energy use	15
Consumption per head,			
kg oil equivalent	2,436		

Inflation and finance

		av. ann. increase 2001–06	
Consumer price			
inflation 2007	2.4%	Narrow money (M1)	24.0%
Av. ann. inflation 2002–07	2.0%	Broad money	7.6%
Money market rate, 2007	4.42%	Household saving rate, 2007	7.4%

Exchange rates

	end 2007		December 2007
Zl per $	2.44	Effective rates	2000 = 100
Zl per SDR	3.85	– nominal	119.30
Zl per €	3.59	– real	120.80

Trade

Principal exports	$bn fob	Principal imports	$bn cif
Machinery & transport equipment	43.9	Machinery & transport equipment	44.8
Manufactured goods	25.2	Manufactured goods	26.0
Other manufactured goods	14.6	Chemicals	16.8
Agric. products & foodstuffs	9.2	Mineral fuels	13.0
Total incl. others	109.6	Total incl. others	125.6

Main export destinations	% of total	Main origins of imports	% of total
Germany	27.1	Germany	24.0
Italy	6.5	Russia	9.7
France	6.2	Italy	6.8
United Kingdom	5.7	China	6.1

Balance of payments, reserves and debt, $bn

Visible exports fob	117.5	Change in reserves	5.9
Visible imports fob	-124.5	Level of reserves	
Trade balance	-7.0	end Dec.	48.5
Invisibles inflows	24.7	No. months of import cover	3.6
Invisibles outflows	-37.0	Official gold holdings, m oz	3.3
Net transfers	8.2	Foreign debt	125.8
Current account balance	-11.1	– as % of GDP	41
– as % of GDP	-3.3	– as % of total exports	97
Capital balance	14.2	Debt service ratio	25
Overall balance	2.5		

Health and education

Health spending, % of GDP	6.2	Education spending, % of GDP	5.5
Doctors per 1,000 pop.	2.0	Enrolment, %: primary	98
Hospital beds per 1,000 pop.	5.3	secondary	100
Improved-water source access, % of pop.	...	tertiary	66

Society

No. of households	13.7m	Colour TVs per 100 households	98.5
Av. no. per household	2.8	Telephone lines per 100 pop.	29.8
Marriages per 1,000 pop.	5.1	Mobile telephone subscribers	
Divorces per 1,000 pop.	1.5	per 100 pop.	95.5
Cost of living, Dec. 2007		Computers per 100 pop.	24.2
New York = 100	91	Internet hosts per 1,000 pop.	185.3

PORTUGAL

Area	88,940 sq km	Capital	Lisbon
Arable as % of total land	14	Currency	Euro (€)

People

Population	10.5m	Life expectancy:	men	75.0 yrs
Pop. per sq km	118.0		women	81.2 yrs
Av. ann. growth		Adult literacy		93.8%
in pop. 2010–15	0.15%	Fertility rate (per woman)		1.5
Pop. under 15	15.7%	Urban population		55.6%
Pop. over 60	22.1%			per 1,000 pop.
No. of men per 100 women	94	Crude birth rate		10
Human Development Index	89.7	Crude death rate		10.6

The economy

GDP	€155bn	GDP per head	$18,550
GDP	$195bn	GDP per head in purchasing	
Av. ann. growth in real		power parity (USA=100)	47.3
GDP 1997–2007	2.2%	Economic freedom index	64.3

Origins of GDP		Components of GDP	
	% of total		% of total
Agriculture	8.3	Private consumption	65.0
Industry, of which:	25.0	Public consumption	20.7
manufacturing	...	Investment	21.6
Services	66.2	Exports	31.2
		Imports	-39.2

Structure of employment

	% of total		% of labour force
Agriculture	12	Unemployed 2006	7.7
Industry	31	Av. ann. rate 1995–2006	6.0
Services	57		

Energy

	m TOE		
Total output	3.6	Net energy imports as %	
Total consumption	27.2	of energy use	87
Consumption per head,			
kg oil equivalent	2,575		

Inflation and finance

Consumer price		av. ann. increase 2001–06	
inflation 2007	2.8%	Euro area:	
Av. ann. inflation 2002–07	2.7%	Narrow money (M1)	10.5%
Deposit rate, h'holds, 2007	2.80%	Broad money	7.4%
		Household saving rate, 2007	7.5%

Exchange rates

	end 2007		December 2007
€ per $	0.68	Effective rates	2000 = 100
€ per SDR	1.07	– nominal	108.4
		– real	114.0

Trade

Principal exports		Principal imports	
	$bn fob		*$bn cif*
Machinery & transport equip.	14.3	Machinery & transport equip.	21.1
Food, drink & tobacco	3.4	Fuels & minerals	10.3
Chemicals & related products	3.0	Chemicals & related products	7.6
Fuels & minerals	2.4	Food, drink & tobacco	7.1
Total incl. others	**43.3**	Total incl. others	**66.7**

Main export destinations		Main origins of imports	
	% of total		*% of total*
Spain	27.4	Spain	30.5
Germany	13.1	Germany	13.8
France	12.4	France	8.4
United Kingdom	7.1	Italy	5.8
Italy	4.1	Netherlands	4.5
EU25	77.2	EU25	75.5

Balance of payments, reserves and debt, $bn

Visible exports fob	43.6	Overall balance	-2.4
Visible imports fob	-64.5	Change in reserves	-0.5
Trade balance	-20.9	Level of reserves	
Invisibles inflows	30.1	end Dec.	9.9
Invisibles outflows	-36.9	No. months of import cover	1.2
Net transfers	3.1	Official gold holdings, m oz	12.3
Current account balance	-18.3	Aid given	0.40
– as % of GDP	-9.4	– as % of GDP	0.20
Capital balance	14.3		

Health and education

Health spending, % of GDP	10.2	Education spending, % of GDP	5.5
Doctors per 1,000 pop.	3.4	Enrolment, %: primary	115
Hospital beds per 1,000 pop.	3.7	secondary[a]	97
Improved-water source access, % of pop.	...	tertiary	55

Society

No. of households	3.9m	Colour TVs per 100 households	98.7
Av. no. per household	2.7	Telephone lines per 100 pop.	40.2
Marriages per 1,000 pop.	5.0	Mobile telephone subscribers	
Divorces per 1,000 pop.	2.4	per 100 pop.	116.0
Cost of living, Dec. 2007		Computers per 100 pop.	13.3
New York = 100	89	Internet hosts per 1,000 pop.	156.5

a Includes training for unemployed.

ROMANIA

Area	237,500 sq km	Capital	Bucharest
Arable as % of total land	40	Currency	Leu (RON)

People

Population	21.6m	Life expectancy:	men	69.0 yrs
Pop. per sq km	90.9		women	76.1 yrs
Av. ann. growth		Adult literacy		97.3%
in pop. 2010–15	-0.53%	Fertility rate (per woman)		1.3
Pop. under 15	15.7%	Urban population		54.7%
Pop. over 60	19.3%			per 1,000 pop.
No. of men per 100 women	95	Crude birth rate		10
Human Development Index	81.3	Crude death rate		12.4

The economy

GDP	RON342bn	GDP per head	$5,630
GDP	$122bn	GDP per head in purchasing	
Av. ann. growth in real		power parity (USA=100)	23.7
GDP 1996–2006	4.2%	Economic freedom index	61.5

Origins of GDP		Components of GDP	
	% of total		% of total
Agriculture	10.0	Private consumption	69.9
Industry, of which:	35.9	Public consumption	18.0
manufacturing	27.5	Investment	24.6
Services	55.2	Exports	32.4
		Imports	-44.5

Structure of employment

	% of total		% of labour force
Agriculture	32	Unemployed 2006	7.3
Industry	30	Av. ann. rate 1995–2006	7.1
Services	38		

Energy

	m TOE		
Total output	27.9	Net energy imports as %	
Total consumption	38.3	of energy use	27
Consumption per head,			
kg oil equivalent	1,772		

Inflation and finance

		av. ann. increase 2001–06	
Consumer price			
inflation 2007	4.8%	Narrow money (M1)	40.6%
Av. ann. inflation 2002–07	9.5%	Broad money	33.5%
Money market rate, 2007	7.55%		

Exchange rates

	end 2007		December 2007
RON per $	2.46	Effective rates	2000 = 100
RON per SDR	3.88	– nominal	61.78
RON per €	3.62	– real	136.56

Trade

Principal exports		Principal imports	
	$bn fob		*$bn cif*
Machinery & equipment	6.6	Machinery & equipment	12.3
Textiles	5.3	Fuels & minerals	7.6
Basic metals & products	4.9	Textiles & footwear	4.3
Minerals & fuels	3.4	Chemicals	3.9
Total incl. others	**32.3**	Total incl. others	**51.1**

Main export destinations		Main origins of imports	
	% of total		*% of total*
Italy	17.9	Germany	15.2
Germany	15.7	Italy	14.6
France	7.5	China	7.9
EU25	67.7	EU25	62.6

Balance of payments, reserves and debt, $bn

Visible exports fob	32.3	Change in reserves	8.6
Visible imports fob	-47.2	Level of reserves	
Trade balance	-14.8	end Dec.	30.2
Invisibles inflows	9.2	No. months of import cover	6.0
Invisibles outflows	-13.3	Official gold holdings, m oz	3.4
Net transfers	6.1	Foreign debt	55.1
Current account balance	-12.8	– as % of GDP	58
– as % of GDP	-10.5	– as % of total exports	141
Capital balance	18.9	Debt service ratio	18
Overall balance	6.6		

Health and education

Health spending, % of GDP	5.5	Education spending, % of GDP	3.5
Doctors per 1,000 pop.	1.9	Enrolment, %: primary	105
Hospital beds per 1,000 pop.	6.6	secondary	86
Improved-water source access,		tertiary	52
% of pop.	57		

Society

No. of households	7.6m	Colour TVs per 100 households	86.3
Av. no. per household	2.8	Telephone lines per 100 pop.	19.4
Marriages per 1,000 pop.	6.2	Mobile telephone subscribers	
Divorces per 1,000 pop.	1.7	per 100 pop.	80.5
Cost of living, Dec. 2007		Computers per 100 pop.	12.9
New York = 100	74	Internet hosts per 1,000 pop.	93.3

RUSSIA

Area	17,075,400 sq km	Capital	Moscow
Arable as % of total land	7	Currency	Rouble (Rb)

People

Population	142.5m	Life expectancy:	men	59.0 yrs
Pop. per sq km	8.3		women	72.6 yrs
Av. ann. growth		Adult literacy		99.4%
in pop. 2010–15	-0.56%	Fertility rate (per woman)		1.4
Pop. under 15	15.1%	Urban population		73.3%
Pop. over 60	17.1%			per 1,000 pop.
No. of men per 100 women	86	Crude birth rate		10
Human Development Index	80.2	Crude death rate		16.2

The economy

GDP	Rb26,781bn	GDP per head	$6,930
GDP	$987bn	GDP per head in purchasing	
Av. ann. growth in real		power parity (USA=100)	29.8
GDP 1997–2007	6.3%	Economic freedom index	49.9

Origins of GDP		Components of GDP	
	% of total		% of total
Agriculture	4.9	Private consumption	48.5
Industry, of which:	39.3	Public consumption	17.0
manufacturing	19.4	Investment	18.4
Services	55.8	Exports	33.8
		Imports	-21.1

Structure of employment

	% of total		% of labour force
Agriculture	10	Unemployed 2006	7.2
Industry	30	Av. ann. rate 1995–2006	9.6
Services	60		

Energy

	m TOE		
Total output	1,184.9	Net energy imports as %	
Total consumption	646.7	of energy use	-83
Consumption per head,			
kg oil equivalent	4,517		

Inflation and finance

		av. ann. increase 2001–06	
Consumer price			
inflation 2007	9.0%	Narrow money (M1)	36.5%
Av. ann. inflation 2002–07	11.2%	Broad money	36.5%
Money market rate, 2007	4.43%		

Exchange rates

	end 2007		December 2007
Rb per $	24.55	Effective rates	2000 = 100
Rb per SDR	38.79	– nominal	100.24
Rb per 7	36.14	– real	176.80

Trade

Principal exports		Principal imports	
	$bn fob		*$bn fob*
Fuels	196.8	Machinery & equipment	65.6
Metals	41.8	Food & drink	21.8
Machinery & equipment	17.5	Chemicals	21.6
Chemicals	16.9	Metals	9.6
Total incl. others	**303.6**	Total incl. others	**164.3**

Main export destinations		Main origins of imports	
	% of total		*% of total*
Netherlands	11.8	Germany	11.2
Italy	8.3	China	7.8
Germany	8.1	Ukraine	5.6
China	5.2	United States	3.9

Balance of payments, reserves and debt, $bn

Visible exports fob	303.9	Change in reserves	121.5
Visible imports fob	-164.7	Level of reserves	
Trade balance	139.2	end Dec.	303.8
Invisibles inflows	60.4	No. months of import cover	13.6
Invisibles outflows	-103.9	Official gold holdings, m oz	12.9
Net transfers	-1.5	Foreign debt	251.1
Current account balance	94.3	– as % of GDP	34
– as % of GDP	9.6	– as % of total exports	88
Capital balance	5.8	Debt service ratio	14
Overall balance	107.5		

Health and education

Health spending, % of GDP	5.2	Education spending, % of GDP	3.8
Doctors per 1,000 pop.	4.3	Enrolment, %: primary	96
Hospital beds per 1,000 pop.	9.7	secondary	84
Improved-water source access,		tertiary	72
% of pop.	97		

Society

No. of households	53.0m	Colour TVs per 100 households	93.9
Av. no. per household	2.7	Telephone lines per 100 pop.	30.8
Marriages per 1,000 pop.	7.6	Mobile telephone subscribers	
Divorces per 1,000 pop.	4.1	per 100 pop.	105.7
Cost of living, Dec. 2007		Computers per 100 pop.	12.2
New York = 100	107	Internet hosts per 1,000 pop.	25.1

SAUDI ARABIA

Area	2,200,000 sq km	Capital	Riyadh
Arable as % of total land	2	Currency	Riyal (SR)

People

Population	25.2m	Life expectancy: men	70.9 yrs
Pop. per sq km	11.6	women	75.3 yrs
Av. ann. growth		Adult literacy	85.0%
in pop. 2010–15	2.05%	Fertility rate (per woman)	3.0
Pop. under 15	34.5%	Urban population	88.5%
Pop. over 60	4.2%		per 1,000 pop.
No. of men per 100 women	122	Crude birth rate	30
Human Development Index	81.2	Crude death rate	3.7

The economy

GDP	SR1,308bn	GDP per head	$13,850
GDP	$349bn	GDP per head in purchasing	
Av. ann. growth in real		power parity (USA=100)	50.7
GDP 1996–2006	3.9%	Economic freedom index	62.8

Origins of GDP		Components of GDP	
	% of total		% of total
Agriculture	3.0	Private consumption	25.5
Industry, of which:	64.7	Public consumption	25.2
manufacturing	9.4	Investment	17.0
Services	32.3	Exports	62.2
		Imports	-30.7

Structure of employment

	% of total		% of labour force
Agriculture	7	Unemployed 2006	6.3
Industry	21	Av. ann. rate 1995–2006	4.6
Services	72		

Energy

	m TOE		
Total output	576.7	Net energy imports as %	
Total consumption	140.3	of energy use	-311
Consumption per head,			
kg oil equivalent	6,068		

Inflation and finance

Consumer price		av. ann. increase 2001–06	
inflation 2007	4.2%	Narrow money (M1)	11.7%
Av. ann. inflation 2002–07	1.6%	Broad money	14.8%
Deposit rate, 2007	4.90%		

Exchange rates

	end 2007		December 2007
SR per $	3.75	Effective rates	2000 = 100
SR per SDR	5.93	– nominal	84.00
SRE per €	5.52	– real	77.60

Trade

Principal exports		Principal imports	
	$bn fob		*$bn cif*
Crude oil & refined		Machinery & transport	
petroleum	137.2	equipment	29.9
Oil products	24.8	Foodstuffs	8.3
Total incl. others	**209.5**	Total incl. others	**42.0**

Main export destinations		Main origins of imports	
	% of total		*% of total*
Japan	17.7	United States	12.3
United States	15.9	Germany	8.6
South Korea	9.1	Japan	7.9
China	7.2	United Kingdom	7.3

Balance of payments, reserves and aid, $bn

Visible exports fob	211.3	Overall balance	0.9
Visible imports fob	-63.9	Change in reserves	1.6
Trade balance	147.4	Level of reserves	
Invisibles inflows	17.7	end Dec.	30.4
Invisibles outflows	-50.3	No. months of import cover	3.2
Net transfers	-15.7	Official gold holdings, m oz	4.6
Current account balance	99.1	Aid given	2.09
– as % of GDP	28.4	– as % of GDP	0.60
Capital balance	-98.2		

Health and education

Health spending, % of GDP	3.4	Education spending, % of GDP	6.8
Doctors per 1,000 pop.	0.6	Enrolment, %: primary	68
Hospital beds per 1,000 pop.	2.3	secondary	68
Improved-water source access,		tertiary	29
% of pop.	96		

Society

No. of households	4.3m	Colour TVs per 100 households	97.2
Av. no. per household	5.8	Telephone lines per 100 pop.	15.7
Marriages per 1,000 pop.	4.2	Mobile telephone subscribers	
Divorces per 1,000 pop.	0.9	per 100 pop.	78.1
Cost of living, Dec. 2007		Computers per 100 pop.	13.6
New York = 100	64	Internet hosts per 1,000 pop.	5.1

SINGAPORE

Area	639 sq km	Capital	Singapore
Arable as % of total land	1	Currency	Singapore dollar (S$)

People

Population	4.4m	Life expectancy: men		78.0 yrs
Pop. per sq km	6,885.8	women		81.9 yrs
Av. ann. growth		Adult literacy		92.5%
in pop. 2010–15	0.92%	Fertility rate (per woman)		1.3
Pop. under 15	19.5%	Urban population		100.0%
Pop. over 60	12.3%			per 1,000 pop.
No. of men per 100 women	101	Crude birth rate		10
Human Development Index	92.2	Crude death rate		5.3

The economy

GDP	S$210bn	GDP per head	$30,040
GDP	$132bn	GDP per head in purchasing	
Av. ann. growth in real		power parity (USA=100)	101.7
GDP 1997–2007	5.8%	Economic freedom index	87.4

Origins of GDP		Components of GDP	
	% of total		% of total
Agriculture	0	Private consumption	40.1
Industry, of which:	34	Public consumption	11.2
manufacturing	28	Investment	22.7
Services	66	Exports	246.2
		Imports	-216.3

Structure of employment

	% of total		% of labour force
Agriculture	0	Unemployed 2006	4.5
Industry	30	Av. ann. rate 1995–2006	3.9
Services	70		

Energy

	m TOE		
Total output	0.0	Net energy imports as %	
Total consumption	30.1	of energy use	100
Consumption per head,			
kg oil equivalent	6,933		

Inflation and finance

		av. ann. increase 2001–06	
Consumer price			
inflation 2007	2.1%	Narrow money (M1)	7.7%
Av. ann. inflation 2002–07	1.1%	Broad money	7.9%
Money market rate, 2007	2.72%		

Exchange rates

	end 2007		December 2007
S$ per $	1.44	Effective rates	2000 = 100
S$ per SDR	2.28	– nominal	106.00
S$ per 7	2.12	– real	97.90

Trade

Principal exports		Principal imports	
	$bn fob		$bn cif
Machinery & equipment	156.9	Machinery & equipment	130.5
Mineral fuels	35.1	Petroleum	20.4
Chemicals	30.9	Manufactured products	17.6
Manufactured products	11.6	Chemicals	14.3
Food	2.5	Food	4.3
Total incl. others	**271.9**	Total incl. others	**238.8**

Main export destinations		Main origins of imports	
	% of total		% of total
Malaysia	13.1	Malaysia	13.0
Hong Kong	10.1	United States	12.5
United States	9.9	China	10.7
China	9.7	Japan	8.3
Japan	5.5	Saudi Arabia	3.9
Thailand	4.2	Thailand	3.7
South Korea	3.5	Hong Kong	1.7

Balance of payments, reserves and debt, $bn

Visible exports fob	275.0	Change in reserves	20.1
Visible imports fob	-230.2	Level of reserves	
Trade balance	44.7	end Dec.	136.3
Invisibles inflows	89.4	No. months of import cover	5.0
Invisibles outflows	-96.5	Official gold holdings, m oz	...
Net transfers	-1.4	Foreign debt	24.0
Current account balance	36.3	– as % of GDP	18
– as % of GDP	27.5	– as % of total exports	7
Capital balance	-21.0	Debt service ratio	1
Overall balance	17.0		

Health and education

Health spending, % of GDP	3.5	Education spending, % of GDP	...
Doctors per 1,000 pop.	1.6	Enrolment, %: primary	80
Hospital beds per 1,000 pop.	2.8	secondary	...
Improved-water source access,		tertiary	44
% of pop.	100		

Society

No. of households	1.0m	Colour TVs per 100 households	99.4
Av. no. per household	4.3	Telephone lines per 100 pop.	42.3
Marriages per 1,000 pop.	5.2	Mobile telephone subscribers	
Divorces per 1,000 pop.	1.6	per 100 pop.	109.3
Cost of living, Dec. 2007		Computers per 100 pop.	68.2
New York = 100	112	Internet hosts per 1,000 pop.	193.8

SLOVAKIA

Area	49,035 sq km	Capital	Bratislava
Arable as % of total land	29	Currency	Koruna (Sk)

People

Population	5.4m	Life expectancy: men		70.7 yrs
Pop. per sq km	110.1	women		78.5 yrs
Av. ann. growth		Adult literacy		...
in pop. 2010–15	-0.02%	Fertility rate (per woman)		1.3
Pop. under 15	16.8%	Urban population		58.0%
Pop. over 60	16.1%			per 1,000 pop.
No. of men per 100 women	94	Crude birth rate		10.0
Human Development Index	86.3	Crude death rate		10.0

The economy

GDP	Sk1,636bn	GDP per head	$10,190
GDP	$55.0bn	GDP per head in purchasing	
Av. ann. growth in real		power parity (USA=100)	40.3
GDP 1997–2007	5.5%	Economic freedom index	68.7

Origins of GDP		Components of GDP	
	% of total		% of total
Agriculture	3.6	Private consumption	55.9
Industry, of which:	31.6	Public consumption	19.0
Manufacturing	19.8	Investment	26.3
Services	64.9	Exports	84.4
		Imports	-88.2

Structure of employment

	% of total		% of labour force
Agriculture	5	Unemployed 2006	13.3
Industry	39	Av. ann. rate 1995–2006	15.6
Services	56		

Energy

	m TOE		
Total output	6.6	Net energy imports as %	
Total consumption	18.8	of energy use	65
Consumption per head,			
kg oil equivalent	3,496		

Inflation and finance

Consumer price		av. ann. increase 2001–06	
inflation 2007	2.8%	Narrow money	18.6%
Av. ann. inflation 2002–07	5.2%	Broad money	7.7%
Money market rate, 2007	4.25%	Household saving rate, 2007	9.9%

Exchange rates

	end 2007		December 2007
Sk per $	22.87	Effective rates	2000 = 100
Sk per SDR	36.14	– nominal	131.22
Sk per €	33.67	– real	161.20

Trade

Principal exports		Principal imports	
	$bn fob		*$bn fob*
Machinery & transport		Machinery & transport	
equipment	20.4	equipment	17.2
Semi-manufactures	9.9	Semi-manufactures	7.6
Other manufactured goods	3.8	Fuels	6.0
Chemicals	2.5	Chemicals	4.3
Total incl. others	**41.9**	Total incl. others	**44.3**

Main export destinations		Main origins of imports	
	% of total		*% of total*
Germany	21.4	Germany	19.8
Czech Republic	12.4	Czech Republic	11.4
Italy	6.4	Russia	9.4
Poland	6.2	Hungary	3.1
EU25	86.7	EU25	68.9

Balance of payments, reserves and debt, $bn

Visible exports fob	41.5	Change in reserves	-2.1
Visible imports fob	44.0	Level of reserves	
Trade balance	-2.5	end Dec.	13.4
Invisibles inflows	7.4	No. months of import cover	3.0
Invisibles outflows	-8.7	Official gold holdings, m oz	1.1
Net transfers	-0.1	Foreign debt	27.1
Current account balance	-3.9	– as % of GDP	58
– as % of GDP	-7.1	– as % of total exports	67
Capital balance	1.0	Debt service ratio	8
Overall balance	2.6		

Health and education

Health spending, % of GDP	7.0	Education spending, % of GDP	3.9
Doctors per 1,000 pop.	3.1	Enrolment, %: primary	100
Hospital beds per 1,000 pop.	6.9	secondary	94
Improved-water source access,		tertiary	45
% of pop.	100		

Society

No. of households	2.2m	Colour TVs per 100 households	96.1
Av. no. per household	2.5	Telephone lines per 100 pop.	21.6
Marriages per 1,000 pop.	5.0	Mobile telephone subscribers	
Divorces per 1,000 pop.	2.2	per 100 pop.	90.6
Cost of living, Dec. 2007		Computers per 100 pop.	35.8
New York = 100	...	Internet hosts per 1,000 pop.	128.8

SLOVENIA

Area	20,253 sq km	Capital	Ljubljana
Arable as % of total land	9	Currency	Tolars (SIT)/Euro (€)

People

Population	2.0m	Life expectancy: men	74.1 yrs
Pop. per sq km	98.8	women	81.5 yrs
Av. ann. growth		Adult literacy	99.7%
in pop. 2010–15	-0.08%	Fertility rate (per woman)	1.3
Pop. under 15	14.1%	Urban population	50.8%
Pop. over 60	20.5%		per 1,000 pop.
No. of men per 100 women	95	Crude birth rate	9.0
Human Development Index	91.7	Crude death rate	9.9

The economy

GDP	SIT7,126bn	GDP per head	$18,650
GDP	$37.3bn	GDP per head in purchasing	
Av. ann. growth in real		power parity (USA=100)	55.4
GDP 1997–2007	4.8%	Economic freedom index	60.6

Origins of GDP		Components of GDP	
	% of total		% of total
Agriculture	2.3	Private consumption	53.4
Industry, of which:	33.7	Public consumption	19.2
manufacturing	24.0	Investment	26.1
Services	63.5	Exports	67.4
		Imports	-68.4

Structure of employment

	% of total		% of labour force
Agriculture	9	Unemployed 2006	5.9
Industry	37	Av. ann. rate 1995–2006	6.7
Services	54		

Energy

	m TOE		
Total output	3.4	Net energy imports as %	
Total consumption	7.3	of energy use	53
Consumption per head,			
kg oil equivalent	3,657		

Inflation and finance

Consumer price		av. ann. increase 2000–05	
inflation 2007	3.6%	Narrow money (M1)	15.9%
Av. ann. inflation 2002–07	3.6%	Broad money	11.9%
Money market rate, 2007	4.08%		

Exchange rates

	end 2007		December 2007
€ per $	0.68	Effective rates	2000 = 100
€ per SDR	1.07	– nominal	...
		– real	...

Trade

Principal exports		Principal imports	
	$bn fob		*$bn fob*
Machinery & transport equip.	8.0	Machinery & transport equip.	7.5
Manufactures	5.4	Manufactures	5.3
Chemicals	2.9	Chemicals	2.8
Miscellaneous manufactures	2.9	Miscellaneous manufactures	2.2
Total incl. others	**21.0**	Total incl. others	**23.0**

Main export destinations		Main origins of imports	
	% of total		*% of total*
Germany	21.2	Germany	20.5
Italy	13.8	Italy	18.7
Croatia	9.7	Austria	12.3
Austria	9.3	France	6.2
France	6.9	Croatia	4.8
EU25	68.4	EU25	80.2

Balance of payments, reserves and debt, $bn

Visible exports fob	21.4	Change in reserves	-1.0
Visible imports fob	-22.9	Level of reserves	
Trade balance	-1.5	end Dec.	7.1
Invisibles inflows	5.5	No. months of import cover	3.1
Invisibles outflows	-4.9	Official gold holdings, m oz	0.2
Net transfers	-0.2	Foreign debt	21.0
Current account balance	-1.1	– as % of GDP	56
– as % of GDP	-2.9	– as % of total exports	80
Capital balance	-0.3	Debt service ratio	19
Overall balance	-1.7		

Health and education

Health spending, % of GDP	8.5	Education spending, % of GDP	6.0
Doctors per 1,000 pop.	2.4	Enrolment, %: primary	100
Hospital beds per 1,000 pop.	4.8	secondary	95
Improved-water source access,		tertiary	83
% of pop.	...		

Society

No. of households	0.7m	Colour TVs per 100 households	96.7
Av. no. per household	2.9	Telephone lines per 100 pop.	42.6
Marriages per 1,000 pop.	3.2	Mobile telephone subscribers	
Divorces per 1,000 pop.	1.2	per 100 pop.	92.6
Cost of living, Dec. 2007		Computers per 100 pop.	40.4
New York = 100	...	Internet hosts per 1,000 pop.	34.7

SOUTH AFRICA

Area	1,225,815 sq km	Capital	Pretoria
Arable as % of total land	12	Currency	Rand (R)

People

Population	47.6m	Life expectancy: men	48.8 yrs
Pop. per sq km	38.8	women	49.7 yrs
Av. ann. growth		Adult literacy	...
in pop. 2010–15	0.40%	Fertility rate (per woman)	2.5
Pop. under 15	32.1%	Urban population	57.9%
Pop. over 60	6.7%		per 1,000 pop.
No. of men per 100 women	97	Crude birth rate	23
Human Development Index	67.4	Crude death rate	17.0

The economy

GDP	R1,727bn	GDP per head	$5,360
GDP	$255bn	GDP per head in purchasing	
Av. ann. growth in real		power parity (USA=100)	20.7
GDP 1997–2007	4.1%	Economic freedom index	63.2

Origins of GDP		**Components of GDP**	
	% of total		% of total
Agriculture	2.8	Private consumption	62.5
Industry, of which:	31.2	Public consumption	19.5
manufacturing	18.4	Investment	18.6
Services	66.0	Exports	29.6
		Imports	-32.9

Structure of employment

	% of total		% of labour force
Agriculture	10	Unemployed 2006	25.5
Industry	24	Av. ann. rate 1995–2006	24.5
Services	66		

Energy

	m TOE		
Total output	158.6	Net energy imports as %	
Total consumption	127.6	of energy use	-24
Consumption per head,			
kg oil equivalent	2,722		

Inflation and finance

Consumer price		av. ann. increase 2001–06	
inflation 2007	7.1%	Narrow money (M1)	14.0%
Av. ann. inflation 2002–07	4.5%	Broad money	14.3%
Money market rate, 2007	9.22%		

Exchange rates

	end 2007		December 2007
R per $	6.81	Effective rates	2000 = 100
R per SDR	10.76	– nominal	103.00
R per €	10.03	– real	98.90

Trade

Principal exports		Principal imports	
	$bn fob		*$bn cif*
Platinum	7.4	Cars & other components	8.8
Gold	4.7	Petrochemicals	8.7
Coal	2.8	Petroleum oils & other	2.5
Cars & other components	2.7	Telecoms components	2.5
Ferro-alloys	2.5		
Total incl. others	**58.3**	Total incl. others	**67.7**

Main export destinations		Main origins of imports	
	% of total		*% of total*
Japan	10.7	Germany	13.9
United States	10.6	China	11.1
United Kingdom	7.9	United States	8.4
Germany	6.8	Japan	7.3

Balance of payments, reserves and debt, $bn

Visible exports fob	63.8	Change in reserves	5.0
Visible imports fob	-69.9	Level of reserves	
Trade balance	-6.1	end Dec.	25.6
Invisibles inflows	18.0	No. months of import cover	3.2
Invisibles outflows	-25.5	Official gold holdings, m oz	4.0
Net transfers	-2.8	Foreign debt	35.6
Current account balance	-16.5	– as % of GDP	15
– as % of GDP	-6.5	– as % of total exports	51
Capital balance	14.7	Debt service ratio	7
Overall balance	3.7		

Health and education

Health spending, % of GDP	8.9	Education spending, % of GDP	5.4
Doctors per 1,000 pop.	0.8	Enrolment, %: primary	106
Hospital beds per 1,000 pop.	...	secondary	95
Improved-water source access,		tertiary	15
% of pop.	88		

Society

No. of households	12.8m	Colour TVs per 100 households	60.9
Av. no. per household	3.9	Telephone lines per 100 pop.	10.0
Marriages per 1,000 pop.	3.8	Mobile telephone subscribers	
Divorces per 1,000 pop.	1.0	per 100 pop.	83.3
Cost of living, Dec. 2007		Computers per 100 pop.	8.5
New York = 100	68	Internet hosts per 1,000 pop.	25.1

SOUTH KOREA

Area	99,274 sq km	Capital	Seoul
Arable as % of total land	16	Currency	Won (W)

People

Population	48.0m	Life expectancy: men	75.0 yrs
Pop. per sq km	483.5	women	82.2 yrs
Av. ann. growth		Adult literacy	...
in pop. 2010–15	0.18%	Fertility rate (per woman)	1.2
Pop. under 15	18.6%	Urban population	80.8%
Pop. over 60	13.7%		per 1,000 pop.
No. of men per 100 women	100	Crude birth rate	9
Human Development Index	92.1	Crude death rate	5.9

The economy

GDP	W847trn	GDP per head	$18,500
GDP	$888bn	GDP per head in purchasing	
Av. ann. growth in real		power parity (USA=100)	52.3
GDP 1997–2007	4.8%	Economic freedom index	67.9

Origins of GDP		Components of GDP	
	% of total		% of total
Agriculture	3.3	Private consumption	53.6
Industry, of which:	39.6	Public consumption	14.8
manufacturing	...	Investment	29.0
Services	57.1	Exports	43.0
		Imports	-42.1

Structure of employment

	% of total		% of labour force
Agriculture	8	Unemployed 2006	3.5
Industry	27	Av. ann. rate 1995–2006	3.7
Services	65		

Energy

	m TOE		
Total output	42.9	Net energy imports as %	
Total consumption	213.8	of energy use	80
Consumption per head,			
kg oil equivalent	4,426		

Inflation and finance

Consumer price		av. ann. increase 2001–06	
inflation 2007	2.5%	Narrow money (M1)	9.9%
Av. ann. inflation 2002–07	2.9%	Broad money	4.8%
Money market rate, 2007	4.77%	Household saving rate, 2007	3.9%

Exchange rates

	end 2007		December 2007
			2000 = 100
W per $	936	Effective rates	
W per SDR	1,479	– nominal	...
W per €	1,378	– real	...

Trade

Principal exports		**Principal imports**	
	$bn fob		*$bn cif*
Information & communications		Crude petroleum	55.9
products	46.0	Machinery & equipment	35.4
Semiconductors	37.4	Semiconductors	28.0
Chemicals	31.2	Chemicals	25.2
Machinery & equipment	29.0		
Total incl. others	**325.5**	Total incl. others	**309.4**

Main export destinations		**Main origins of imports**	
	% of total		*% of total*
China	21.3	Japan	16.8
United States	13.3	China	15.7
Japan	8.2	United States	10.9
Hong Kong	5.8	Germany	3.7

Balance of payments, reserves and debt, $bn

Visible exports fob	331.8	Change in reserves	28.6
Visible imports fob	-302.6	Level of reserves	
Trade balance	29.2	end Dec.	239.1
Invisibles inflows	65.5	No. months of import cover	7.4
Invisibles outflows	-84.8	Official gold holdings, m oz	0.4
Net transfers	-3.8	Foreign debt	187.0
Current account balance	6.1	– as % of GDP	21
– as % of GDP	0.7	– as % of total exports	47
Capital balance	18.6	Debt service ratio	4
Overall balance	22.1		

Health and education

Health spending, % of GDP	5.9	Education spending, % of GDP	4.6
Doctors per 1,000 pop.	1.4	Enrolment, %: primary	105
Hospital beds per 1,000 pop.	7.1	secondary	98
Improved-water source access,		tertiary	93
% of pop.	92		

Society

No. of households	17.6m	Colour TVs per 100 households	99.2
Av. no. per household	2.7	Telephone lines per 100 pop.	49.8
Marriages per 1,000 pop.	6.6	Mobile telephone subscribers	
Divorces per 1,000 pop.	4.1	per 100 pop.	83.8
Cost of living, Dec. 2007		Computers per 100 pop.	53.2
New York = 100	115	Internet hosts per 1,000 pop.	7.1

SPAIN

Area	504,782 sq km	Capital	Madrid
Arable as % of total land	27	Currency	Euro (€)

People

Population	43.4m	Life expectancy:	men	77.7 yrs
Pop. per sq km	86.0		women	84.2 yrs
Av. ann. growth		Adult literacy		...
in pop. 2010–15	0.39%	Fertility rate (per woman)		1.5
Pop. under 15	14.4%	Urban population		76.7%
Pop. over 60	21.7%			per 1,000 pop.
No. of men per 100 women	97	Crude birth rate		11
Human Development Index	94.9	Crude death rate		8.8

The economy

GDP	€976bn	GDP per head	$28,220
GDP	$1,225bn	GDP per head in purchasing	
Av. ann. growth in real		power parity (USA=100)	65.2
GDP 1997–2007	4.3%	Economic freedom index	69.7

Origins of GDP		**Components of GDP**	
	% of total		% of total
Agriculture	3.7	Private consumption	57.4
Industry, of which:	30.4	Public consumption	18.1
manufacturing	...	Investment	30.4
Services	66.0	Exports	26.0
		Imports	-32.2

Structure of employment

	% of total		% of labour force
Agriculture	5	Unemployed 2006	8.5
Industry	30	Av. ann. rate 1995–2006	15.0
Services	65		

Energy

	m TOE		
Total output	30.3	Net energy imports as %	
Total consumption	145.2	of energy use	79
Consumption per head,			
kg oil equivalent	3,346		

Inflation and finance

Consumer price		av. ann. increase 2001–06	
inflation 2007	2.8%	Euro area:	
Av. ann. inflation 2002–07	3.1%	Narrow money (M1)	10.5%
Money market rate, 2007	3.85%	Broad money	7.4%
		Household saving rate, 2007	10.3%

Exchange rates

	end 2007		December 2007
€ per $	0.68	Effective rates	2000 = 100
€ per SDR	1.07	– nominal	109.80
		– real	119.80

Trade

Principal exports		Principal imports	
	$bn fob		*$bn cif*
Raw materials &		Raw materials & intermediate	
intermediate products	104.6	products (excl. fuels)	145.0
Consumer goods	82.3	Consumer goods	88.8
Capital goods	21.6	Energy	49.8
		Capital goods	33.9
Total incl. others	**213.4**	Total incl. others	**326.1**

Main export destinations		Main origins of imports	
	% of total		*% of total*
France	18.7	Germany	14.2
Germany	10.9	France	12.8
Italy	8.8	Italy	8.2
EU25	70.9	EU25	58.8

Balance of payments, reserves and aid, $bn

Visible exports fob	216.5	Overall balance	0.6
Visible imports fob	-317.2	Change in reserves	2.1
Trade balance	-100.7	Level of reserves	
Invisibles inflows	155.4	end Dec.	19.3
Invisibles outflows	-153.9	No. months of import cover	0.5
Net transfers	-7.1	Official gold holdings, m oz	13.4
Current account balance	-106.3	Aid given	3.81
– as % of GDP	-8.7	– as % of GDP	0.31
Capital balance	110.4		

Health and education

Health spending, % of GDP	8.2	Education spending, % of GDP	4.2
Doctors per 1,000 pop.	3.1	Enrolment, %: primary	105
Hospital beds per 1,000 pop.	3.5	secondary	119
Improved-water source access,		tertiary	67
% of pop.	100		

Society

No. of households	15.3m	Colour TVs per 100 households	99.4
Av. no. per household	2.9	Telephone lines per 100 pop.	45.9
Marriages per 1,000 pop.	5.3	Mobile telephone subscribers	
Divorces per 1,000 pop.	1.7	per 100 pop.	106.4
Cost of living, Dec. 2007		Computers per 100 pop.	27.7
New York = 100	108	Internet hosts per 1,000 pop.	71.1

SWEDEN

Area	449,964 sq km	Capital	Stockholm
Arable as % of total land	7	Currency	Swedish krona (Skr)

People

Population	9.1m	Life expectancy: men	78.7 yrs
Pop. per sq km	20.2	women	83.0 yrs
Av. ann. growth		Adult literacy	...
in pop. 2010–15	0.42%	Fertility rate (per woman)	1.8
Pop. under 15	17.4%	Urban population	83.4%
Pop. over 60	23.4%		per 1,000 pop.
No. of men per 100 women	98	Crude birth rate	12
Human Development Index	95.6	Crude death rate	10.1

The economy

GDP	Skr2,832bn	GDP per head	$42,180
GDP	$384bn	GDP per head in purchasing	
Av. ann. growth in real		power parity (USA=100)	77.8
GDP 1997–2007	3.6%	Economic freedom index	70.4

Origins of GDP

	% of total
Agriculture	1.3
Industry, of which:	28.5
manufacturing	...
Services	70.1

Components of GDP

	% of total
Private consumption	47.4
Public consumption	26.3
Investment	18.1
Exports	51.4
Imports	-43.2

Structure of employment

	% of total		% of labour force
Agriculture	2	Unemployed 2006	5.4
Industry	22	Av. ann. rate 1995–2006	6.3
Services	76		

Energy

	m TOE		
Total output	34.8	Net energy imports as %	
Total consumption	52.2	of energy use	33
Consumption per head,			
kg oil equivalent	5,782		

Inflation and finance

Consumer price		av. ann. increase 2001–06	
inflation 2007	2.2%	Narrow money	11.0%
Av. ann. inflation 2002–07	1.3%	Broad money	6.8%
Repurchase rate, 2007	4.00%	Household saving rate, 2007	9.9%

Exchange rates

	end 2007		December 2007
Skr per $	6.41	Effective rates	2000 = 100
Skr per SDR	10.14	– nominal	101.60
Skr per €	9.44	– real	98.80

Trade

Principal exports		Principal imports	
	$bn fob		*$bn cif*
Machinery & transport		Machinery & transport	
equipment	66.5	equipment	50.6
Chemicals	17.3	Fuels & lubricants	16.2
Raw materials	9.0	Chemicals	13.6
Fuels & lubricants	8.9	Food, drink & tobacco	9.3
Total incl. others	**147.9**	Total incl. others	**127.7**

Main export destinations		Main origins of imports	
	% of total		*% of total*
Germany	10.4	Germany	18.4
Norway	9.4	Denmark	9.1
United States	7.6	Norway	8.6
Denmark	7.4	United Kingdom	6.9
United Kingdom	7.1	Netherlands	6.1
EU25	60.9	EU25	71.6

Balance of payments, reserves and aid, $bn

Visible exports fob	148.8	Overall balance	1.3
Visible imports fob	-127.3	Change in reserves	3.2
Trade balance	21.4	Level of reserves	
Invisibles inflows	95.7	end Dec.	28.0
Invisibles outflows	-84.0	No. months of import cover	1.6
Net transfers	-4.6	Official gold holdings, m oz	5.1
Current account balance	28.4	Aid given	3.95
– as % of GDP	7.4	– as % of GDP	1.03
Capital balance	-35.5		

Health and education

Health spending, % of GDP	8.9	Education spending, % of GDP	7.1
Doctors per 1,000 pop.	3.2	Enrolment, %: primary	96
Hospital beds per 1,000 pop.	...	secondary	103
Improved-water source access,		tertiary	79
% of pop.	100		

Society

No. of households	4.2m	Colour TVs per 100 households	97.3
Av. no. per household	2.1	Telephone lines per 100 pop.	59.5
Marriages per 1,000 pop.	4.9	Mobile telephone subscribers	
Divorces per 1,000 pop.	2.2	per 100 pop.	105.9
Cost of living, Dec. 2007		Computers per 100 pop.	83.6
New York = 100	112	Internet hosts per 1,000 pop.	386.1

SWITZERLAND

Area	41,293 sq km	Capital	Berne
Arable as % of total land	10	Currency	Swiss franc (SFr)

People

Population	7.3m	Life expectancy:	men	79.0 yrs
Pop. per sq km	176.8		women	84.2 yrs
Av. ann. growth		Adult literacy		...
in pop. 2010–15	0.35%	Fertility rate (per woman)		1.5
Pop. under 15	16.7%	Urban population		67.5%
Pop. over 60	21.1%			per 1,000 pop.
No. of men per 100 women	95	Crude birth rate		10
Human Development Index	95.5	Crude death rate		8.1

The economy

GDP	SFr477bn	GDP per head	$52,110
GDP	$380bn	GDP per head in purchasing	
Av. ann. growth in real		power parity (USA=100)	84.6
GDP 1997–2007	2.2%	Economic freedom index	79.7

Origins of GDP[a]		**Components of GDP**	
	% of total		% of total
Agriculture	1.0	Private consumption	59.3
Industry, of which:	26.3	Public consumption	11.1
manufacturing	...	Investment	21.3
Services	72.7	Exports	52.4
		Imports	-44.9

Structure of employment

	% of total		% of labour force
Agriculture	4	Unemployed 2006	4.0
Industry	23	Av. ann. rate 1995–2006	3.6
Services	73		

Energy

	m TOE		
Total output	10.9	Net energy imports as %	
Total consumption	27.2	of energy use	60
Consumption per head,			
kg oil equivalent	3,651		

Inflation and finance

		av. ann. increase 2001–06	
Consumer price			
inflation 2007	0.7%	Narrow money (M1)	6.0%
Av. ann. inflation 2002–07	0.9%	Broad money	5.7%
Money market rate, 2007	2.00%	Household saving rate, 2007	10.1%

Exchange rates

	end 2007		December 2007
SFr per $	1.13	Effective rates	2000 = 100
SFr per SDR	1.78	– nominal	106.80
SFr per €	1.66	– real	98.40

Trade

Principal exports		Principal imports	
	$bn		$bn
Chemicals	50.2	Chemicals	28.5
Machinery	30.8	Machinery	25.5
Watches & jewellery	11.0	Motor vehicles	12.4
Metals & metal manufactures	10.7	Precision instruments	9.7
Precision instruments	10.3	Textiles	7.5
Total incl. others	**141.7**	Total incl. others	**132.0**

Main export destinations		Main origins of imports	
	% of total		% of total
Germany	20.2	Germany	33.3
United States	10.3	Italy	11.1
Italy	9.0	France	10.3
France	8.6	Netherlands	5.0
United Kingdom	4.7	Austria	4.5
Spain	3.9	United Kingdom	3.6
EU25	62.1	EU25	81.6

Balance of payments, reserves and aid, $bn

Visible exports fob	167.3	Overall balance	0.4
Visible imports fob	-162.2	Change in reserves	6.9
Trade balance	5.0	Level of reserves	
Invisibles inflows	162.3	end Dec.	64.5
Invisibles outflows	-102.1	No. months of import cover	2.9
Net transfers	-10.3	Official gold holdings, m oz	41.5
Current account balance	54.8	Aid given	1.65
– as % of GDP	14.4	– as % of GDP	0.43
Capital balance	-72.9		

Health and education

Health spending, % of GDP	11.4	Education spending, % of GDP	5.8
Doctors per 1,000 pop.	3.9	Enrolment, %: primary	97
Hospital beds per 1,000 pop.	5.7	secondary	93
Improved-water source access,		tertiary	46
% of pop.	100		

Society

No. of households	3.3m	Colour TVs per 100 households	98.3
Av. no. per household	2.2	Telephone lines per 100 pop.	66.9
Marriages per 1,000 pop.	5.3	Mobile telephone subscribers	
Divorces per 1,000 pop.	2.6	per 100 pop.	99.0
Cost of living, Dec. 2007		Computers per 100 pop.	86.5
New York = 100	117	Internet hosts per 1,000 pop.	453.2

a Latest available.

TAIWAN

Area	36,179 sq km	Capital	Taipei
Arable as % of total land	25	Currency	Taiwan dollar (T$)

People

Population	22.8m	Life expectancy:[a] men	74.7 yrs
Pop. per sq km	630.2	women	80.7 yrs
Av. ann. growth		Adult literacy	96.1%
in pop. 2010–15	0.55%	Fertility rate (per woman)	1.6
Pop. under 15	21.0%	Urban population	...
Pop. over 60	12.1%		per 1,000 pop.
No. of men per 100 women	...	Crude birth rate	13.0
Human Development Index	...	Crude death rate[a]	6.5

The economy

GDP	T$11,859bn	GDP per head	$15,990
GDP	$365bn	GDP per head in purchasing	
Av. ann. growth in real		power parity (USA=100)	72.3
GDP 1997–2007	4.8%	Economic freedom index	71.0

Origins of GDP		Components of GDP	
	% of total		% of total
Agriculture	1.6	Private consumption	60.4
Industry	26.8	Public consumption	12.6
Services	71.5	Investment	21.2
		Exports	69.8
		Imports	-64.2

Structure of employment

	% of total		% of labour force
Agriculture	6	Unemployed 2005	4.1
Industry	37	Av. ann. rate 1995–2005	3.3
Services	58		

Energy

	m TOE		
Total output	...	Net energy imports as %	
Total consumption	...	of energy use	...
Consumption per head,			
kg oil equivalent	...		

Inflation and finance

Consumer price		av. ann. increase 2001–06	
inflation 2007	3.3%	Narrow money (M1)	10.3%
Av. ann. inflation 2002–07	1.1%	Broad money	5.5%
Money market rate, 2007	2.31%		

Exchange rates

	end 2007		December 2007
T$ per $	32.43	Effective rates	2000 = 100
T$ per SDR	51.25	– nominal	...
T$ per €	47.42	– real	...

Trade

Principal exports		**Principal imports**	
	$bn fob		*$bn cif*
Machinery & electrical		Machinery & electrical	
equipment	111.6	equipment	72.5
Base metals & manufactures	24.0	Minerals	38.8
Plastics and rubber products	15.9	Metals	23.2
Textiles & clothing	11.8	Chemicals	22.5
Vehicles, aircraft & ships	7.4	Precision instruments, clocks	
		& watches	12.4
Total incl. others	**213.2**	Total incl. others	**201.6**

Main export destinations		**Main origins of imports**	
	% of total		*% of total*
China	24.3	Japan	23.0
Hong Kong	17.5	China	12.3
United States	15.2	United States	11.2
Japan	7.6	South Korea	7.4

Balance of payments, reserves and debt, $bn

Visible exports fob	233.8	Change in reserves	12.9
Visible imports fob	−200.4	Level of reserves	
Trade balance	23.4	end Dec.	266.9
Invisibles inflows	46.7	No. months of import cover	13.1
Invisibles outflows	−43.4	Official gold holdings, m oz	0.0
Net transfers	−3.9	Foreign debt	91.0
Current account balance	24.7	– as % of GDP	25
– as % of GDP	6.8	– as % of total exports	34
Capital balance	−19.7	Debt service ratio	4
Overall balance	6.1		

Health and education

Health spending, % of GDP	...	Education spending, % of GDP	...
Doctors per 1,000 pop.	1.7	Enrolment, %: primary	...
Hospital beds per 1,000 pop.	...	secondary	...
Improved-water source access,		tertiary	...
% of pop.	...		

Society

No. of households	7.3m	Colour TVs per 100 households	99.6
Av. no. per household	3.2	Telephone lines per 100 pop.	63.6
Marriages per 1,000 pop.	7.4	Mobile telephone subscribers	
Divorces per 1,000 pop.	3.3	per 100 pop.	102.0
Cost of living, Dec. 2007		Computers per 100 pop.	57.5
New York = 100	84	Internet hosts per 1,000 pop.	224.6

a   2002 estimate.

THAILAND

Area	513,115 sq km	Capital	Bangkok
Arable as % of total land	28	Currency	Baht (Bt)

People

Population	64.8m	Life expectancy: men	66.5 yrs
Pop. per sq km	126.3	women	75.0 yrs
Av. ann. growth		Adult literacy	92.6%
in pop. 2010–15	0.50%	Fertility rate (per woman)	1.9
Pop. under 15	21.7%	Urban population	32.5%
Pop. over 60	11.3%		per 1,000 pop.
No. of men per 100 women	95	Crude birth rate	14
Human Development Index	78.1	Crude death rate	8.5

The economy

GDP	Bt7,816bn	GDP per head	$3,180
GDP	$206bn	GDP per head in purchasing	
Av. ann. growth in real		power parity (USA=100)	17.3
GDP 1996–2006	3.7%	Economic freedom index	63.5

Origins of GDP

	% of total
Agriculture	10.7
Industry, of which:	44.4
manufacturing	35.1
Services	44.8

Components of GDP

	% of total
Private consumption	55.9
Public consumption	11.8
Investment	28.2
Exports	73.5
Imports	-70.0

Structure of employment

	% of total		% of labour force
Agriculture	42	Unemployed 2006	1.2
Industry	20	Av. ann. rate 1995–2006	1.8
Services	38		

Energy

	m TOE		
Total output	54	Net energy imports as %	
Total consumption	100	of energy use	46
Consumption per head,			
kg oil equivalent	1,588		

Inflation and finance

Consumer price		av. ann. increase 2001–06	
inflation 2007	2.2%	Narrow money (M1)	9.7%
Av. ann. inflation 2002–07	3.2%	Broad money	6.5%
Money market rate, 2007	3.75%		

Exchange rates

	end 2007		December 2007
Bt per $	33.72	Effective rates	2000 = 100
Bt per SDR	53.28	– nominal	...
Bt per €	49.64	– real	...

Trade

Principal exports		**Principal imports**	
	$bn fob		*$bn cif*
Machinery & mech. appliances	20.3	Raw materials & intermediates	51.9
Integrated circuits & parts	14.0	Capital goods	33.1
Electrical appliances	10.0	Petroleum & products	25.4
Vehicle parts	9.7	Consumer goods	9.6
Total incl. others	**130.6**	Total incl. others	**130.6**

Main export destinations		**Main origins of imports**	
	% of total		*% of total*
United States	15.1	Japan	19.8
Japan	12.7	United States	10.4
China	9.0	China	6.6
Singapore	6.5	Malaysia	6.5

Balance of payments, reserves and debt, $bn

Visible exports fob	127.9	Change in reserves	14.9
Visible imports fob	-114.0	Level of reserves	
Trade balance	13.9	end Dec.	67.0
Invisibles inflows	28.8	No. months of import cover	5.1
Invisibles outflows	-43.9	Official gold holdings, m oz	2.7
Net transfers	3.4	Foreign debt	55.2
Current account balance	2.2	– as % of GDP	50
– as % of GDP	-1.1	– as % of total exports	40
Capital balance	5.6	Debt service ratio	9
Overall balance	12.7		

Health and education

Health spending, % of GDP	3.5	Education spending, % of GDP	4.2
Doctors per 1,000 pop.	0.3	Enrolment, %: primary	108
Hospital beds per 1,000 pop.	...	secondary	78
Improved-water source access,		tertiary	46
% of pop.	99		

Society

No. of households	17.5m	Colour TVs per 100 households	94.0
Av. no. per household	3.6	Telephone lines per 100 pop.	10.9
Marriages per 1,000 pop.	4.5	Mobile telephone subscribers	
Divorces per 1,000 pop.	1.1	per 100 pop.	62.9
Cost of living, Dec. 2007		Computers per 100 pop.	7.0
New York = 100	74	Internet hosts per 1,000 pop.	16.1

TURKEY

Area	779,452 sq km	Capital	Ankara
Arable as % of total land	31	Currency	Turkish Lira (YTL)

People

Population	74.2m	Life expectancy: men	69.4 yrs
Pop. per sq km	95.2	women	74.3 yrs
Av. ann. growth		Adult literacy	88.7%
in pop. 2010–15	1.10%	Fertility rate (per woman)	2.1
Pop. under 15	28.3%	Urban population	67.3%
Pop. over 60	8.2%		per 1,000 pop.
No. of men per 100 women	101	Crude birth rate	19
Human Development Index	77.5	Crude death rate	5.9

The economy

GDP	YTL576bn	GDP per head	$5,430
GDP	$403bn	GDP per head in purchasing	
Av. ann. growth in real		power parity (USA=100)	19.1
GDP 1997–2007	4.5%	Economic freedom index	60.8

Origins of GDP		Components of GDP	
	% of total		% of total
Agriculture	9.5	Private consumption	70.5
Industry, of which:	28.7	Public consumption	12.3
manufacturing	19.8	Investment	22.3
Services	61.6	Exports	22.7
		Imports	-27.6

Structure of employment

	% of total		% of labour force
Agriculture	30	Unemployed 2006	9.9
Industry	25	Av. ann. rate 1995–2006	8.5
Services	45		

Energy

	m TOE		
Total output	23.6	Net energy imports as %	
Total consumption	85.2	of energy use	72
Consumption per head,			
kg oil equivalent	1,182		

Inflation and finance

Consumer price		av. ann. increase 2001–06	
inflation 2007	8.8%	Narrow money (M1)	44.4%
Av. ann. inflation 2002–07	12.9%	Broad money	24.4%
Money market rate, 2007	17.24%		

Exchange rates

	end 2007		December 2007
YTL per $	1.16	Effective rates	2000 = 100
YTL per SDR	1.84	– nominal	...
YTL per €	1.71	– real	...

Trade

Principal exports		Principal imports	
	$bn fob		*$bn cif*
Textiles & clothing	19.4	Fuels	28.9
Transport equipment	12.7	Chemicals	19.6
Iron & steel	7.8	Mechanical machinery	19.0
Agricultural products	6.3	Transport equipment	13.3
Total incl. others	**85.5**	Total incl. others	**139.6**

Main export destinations		Main origins of imports	
	% of total		*% of total*
Germany	11.3	Russia	12.8
United Kingdom	8.0	Germany	10.6
Italy	7.9	China	6.9
United States	6.0	Italy	6.2
France	5.2	France	5.2
EU25	56.1	EU25	42.2

Balance of payments, reserves and debt, $bn

Visible exports fob	91.9	Change in reserves	10.8
Visible imports fob	-133.2	Level of reserves	
Trade balance	-41.2	end Dec.	63.3
Invisibles inflows	29.0	No. months of import cover	4.9
Invisibles outflows	-22.2	Official gold holdings, m oz	3.7
Net transfers	1.7	Foreign debt	207.9
Current account balance	-32.8	– as % of GDP	61
– as % of GDP	-8.1	– as % of total exports	200
Capital balance	45.8	Debt service ratio	33
Overall balance	10.6		

Health and education

Health spending, % of GDP	7.6	Education spending, % of GDP	4.0
Doctors per 1,000 pop.	1.6	Enrolment, %: primary	94
Hospital beds per 1,000 pop.	2.6	secondary	79
Improved-water source access,		tertiary	35
% of pop.	96		

Society

No. of households	15.7m	Colour TVs per 100 households	92.0
Av. no. per household	4.6	Telephone lines per 100 pop.	25.4
Marriages per 1,000 pop.	6.4	Mobile telephone subscribers	
Divorces per 1,000 pop.	0.8	per 100 pop.	71.0
Cost of living, Dec. 2007		Computers per 100 pop.	5.7
New York = 100	102	Internet hosts per 1,000 pop.	32.7

UKRAINE

Area	603,700 sq km	Capital	Kiev
Arable as % of total land	56	Currency	Hryvnya (UAH)

People

Population	46.0m	Life expectancy: men	62.1 yrs
Pop. per sq km	76.2	women	73.8 yrs
Av. ann. growth		Adult literacy	99.4%
in pop. 2010–15	-0.79%	Fertility rate (per woman)	1.2
Pop. under 15	14.7%	Urban population	67.3%
Pop. over 60	20.6%		per 1,000 pop.
No. of men per 100 women	86	Crude birth rate	10
Human Development Index	78.8	Crude death rate	16.4

The economy

GDP	UAH538bn	GDP per head	$2,310
GDP	$106bn	GDP per head in purchasing	
Av. ann. growth in real		power parity (USA=100)	14.1
GDP 1997–2007	6.2%	Economic freedom index	51.1

Origins of GDP		**Components of GDP**	
	% of total		% of total
Agriculture	10.9	Private consumption	71.3
Industry	33.8	Public consumption	7.3
Services	55.3	Investment	24.0
		Exports	47.2
		Imports	-50.1

Structure of employment

	% of total		% of labour force
Agriculture	19	Unemployed 2006	6.8
Industry	24	Av. ann. rate 1995–2006	9.2
Services	57		

Energy

	m TOE		
Total output	81	Net energy imports as %	
Total consumption	143.2	of energy use	43
Consumption per head,			
kg oil equivalent	3,041		

Inflation and finance

Consumer price		av. ann. increase 2002–06	
inflation 2007	12.8%	Narrow money (M1)	32.2%
Av. ann. inflation 2002–07	9.9%	Broad money	41.7%
Money market rate, 2007	2.27%		

Exchange rates

	end 2007		December 2007
UAH per $	5.05	Effective rates	2000 = 100
UAH per SDR	7.98	– nominal	88.89
UAH per €	7.43	– real	114.59

Trade

Principal exports		Principal imports	
	$bn fob		*$bn cif*
Metals	16.4	Machinery & equipment	13.7
Machinery & equipment	6.9	Fuels, mineral products	13.3
Food & agricultural produce	4.7	Chemicals	3.9
Fuels & mineral products	3.8	Food & agricultural produce	3.2
Chemicals	3.4		
Total incl. others	**38.3**	Total incl. others	**44.7**

Main export destinations		Main origins of imports	
	% of total		*% of total*
Russia	22.6	Russia	31.2
Italy	6.5	Germany	9.7
Turkey	6.2	Turkmenistan	7.9
Germany	3.4	Poland	5.2

Balance of payments, reserves and debt, $bn

Visible exports fob	38.9	Change in reserves	3.0
Visible imports fob	-44.1	Level of reserves	
Trade balance	-5.2	end Dec.	22.4
Invisibles inflows	12.6	No. months of import cover	4.8
Invisibles outflows	-12.2	Official gold holdings, m oz	0.8
Net transfers	3.2	Foreign debt	49.9
Current account balance	-1.6	– as % of GDP	58
– as % of GDP	-1.5	– as % of total exports	106
Capital balance	4.1	Debt service ratio	18
Overall balance	2.4		

Health and education

Health spending, % of GDP	7.0	Education spending, % of GDP	6.3
Doctors per 1,000 pop.	3.1	Enrolment, %: primary	102
Hospital beds per 1,000 pop.	8.7	secondary	93
Improved-water source access,		tertiary	73
% of pop.	96		

Society

No. of households	19.8m	Colour TVs per 100 households	93.1
Av. no. per household	2.3	Telephone lines per 100 pop.	26.8
Marriages per 1,000 pop.	5.9	Mobile telephone subscribers	
Divorces per 1,000 pop.	3.6	per 100 pop.	106.5
Cost of living, Dec. 2007		Computers per 100 pop.	4.5
New York = 100	74	Internet hosts per 1,000 pop.	9.2

UNITED ARAB EMIRATES

Area	83,600 sq km	Capital	Abu Dhabi
Arable as % of total land	1	Currency	Dirham (AED)

People

Population	4.7m	Life expectancy:	men	77.2 yrs
Pop. per sq km	56.2		women	81.5 yrs
Av. ann. growth		Adult literacy		90.4%
in pop. 2010–15	2.13%	Fertility rate (per woman)		2.2
Pop. under 15	19.8%	Urban population		85.5%
Pop. over 60	1.8%			per 1,000 pop.
No. of men per 100 women	210	Crude birth rate		17
Human Development Index	86.8	Crude death rate		1.4

The economy

GDP[a]	AED476bn	GDP per head	$27,600
GDP[a]	$130bn	GDP per head in purchasing	
Av. ann. growth in real		power parity (USA=100)	76.2
GDP 1996–2006	7.3%	Economic freedom index	62.8

Origins of GDP

Components of GDP

	% of total		% of total
Agriculture	2.8	Private consumption	45.9
Industry, of which:	51.0	Public consumption	10.1
manufacturing	13.9	Investment	20.2
Services	46.2	Exports	91.5
		Imports	-67.8

Structure of employment

	% of total		% of labour force
Agriculture	...	Unemployed 2001	2.3
Industry	...	Av. ann. rate 1995–2001	2.1
Services	...		

Energy

			m TOE
Total output	167.9	Net energy imports as %	
Total consumption	46.9	of energy use	-258
Consumption per head,			
kg oil equivalent	11,436		

Inflation and finance

		av. ann. increase 2001–06	
Consumer price			
inflation 2007	14.0%	Narrow money (M1)	24.9%
Av. ann. inflation 2002–07	8.9%	Broad money	20.6%

Exchange rates

	end 2007		December 2007
AED per $	3.67	Effective rates	2000 = 100
AED per SDR	5.80	– nominal	100.00
AED per €	5.40	– real	84.40

Trade

Principal exports		Principal imports[a]	
	$bn fob		*$bn cif*
Crude oil	56.4	Machinery & electrical equip.	18.1
Re-exports	46.0	Precious stones & metals	14.3
Gas	7.1	Transport equipment	10.0
Total incl. others	**142.5**	Total incl. others	**97.9**

Main export destinations		Main origins of imports	
	% of total		*% of total*
Japan	25.9	United States	11.4
South Korea	10.3	China	11.0
Thailand	5.9	India	9.8
India	4.5	Germany	6.2

Balance of payments, reserves and debt, $bn

Visible exports fob	142.6	Change in reserves	6.6
Visible imports fob	-86.1	Level of reserves	
Trade balance	56.4	end Dec.	27.6
Invisibles, net	-12.1	No. months of import cover	3.3
Net transfers	-8.4	Official gold holdings, m oz	0.0
Current account balance	35.9	Foreign debt	46.0
– as % of GDP	27.7	– as % of GDP	28
Capital balance	-16.1	– as % of total exports	28
Overall balance	6.6	Debt service ratio	2

Health and education

Health spending, % of GDP	2.6	Education spending, % of GDP	...
Doctors per 1,000 pop.	1.5	Enrolment, %: primary	104
Hospital beds per 1,000 pop.	2.2	secondary	90
Improved-water source access,		tertiary	23
% of pop.	100		

Society

No. of households	0.7m	Colour TVs per 100 households	99.7
Av. no. per household	6.6	Telephone lines per 100 pop.	28.1
Marriages per 1,000 pop.	3.2	Mobile telephone subscribers	
Divorces per 1,000 pop.	0.9	per 100 pop.	118.5
Cost of living, Dec. 2007		Computers per 100 pop.	25.6
New York = 100	71	Internet hosts per 1,000 pop.	70.1

a 2005

UNITED KINGDOM

Area	242,534 sq km	Capital	London
Arable as % of total land	24	Currency	Pound (£)

People

Population	59.8m	Life expectancy:	men	77.2 yrs
Pop. per sq km	246.6		women	81.6 yrs
Av. ann. growth		Adult literacy		...
in pop. 2010–15	0.41%	Fertility rate (per woman)		1.9
Pop. under 15	18.0%	Urban population		89.2%
Pop. over 60	21.2%			per 1,000 pop.
No. of men per 100 women	96	Crude birth rate		12.0
Human Development Index	94.6	Crude death rate		9.9

The economy

GDP	£1,290bn	GDP per head	$39,750
GDP	$2,377bn	GDP per head in purchasing	
Av. ann. growth in real		power parity (USA=100)	75.3
GDP 1997–2007	3.2%	Economic freedom index	79.5

Origins of GDP		Components of GDP	
	% of total		% of total
Agriculture	0.9	Private consumption	63.2
Industry, of which:	24.0	Public consumption	21.7
manufacturing	...	Investment	18.2
Services	75.1	Exports	25.9
		Imports	-29.5

Structure of employment

	% of total		% of labour force
Agriculture	1	Unemployed 2005	5.0
Industry	22	Av. ann. rate 1995–2005	6.0
Services	77		

Energy

	m TOE		
Total output	204.3	Net energy imports as %	
Total consumption	233.7	of energy use	13
Consumption per head,			
kg oil equivalent	3,884		

Inflation and finance

Consumer price		av. ann. increase 2001–06	
inflation 2007	4.3%	Narrow money (M0)	5.6%
Av. ann. inflation 2002–07	3.2%	Broad money (M4)	9.7%
Money market rate, 2007	5.67%	Household saving rate, 2007	3.3%

Exchange rates

	end 2007		December 2007
£ per $	0.50	Effective rates	2000 = 100
£ per SDR	0.79	– nominal	98.80
£ per €	0.74	– real	104.70

Trade

Principal exports		Principal imports	
	$bn fob		*$bn fob*
Finished manufactured		Finished manufactured	
products	251.8	products	352.9
Semi-manufactured products	119.9	Semi-manufactured products	128.0
Fuels	46.4	Fuels	58.4
Food, drink & tobacco	20.4	Food, drink & tobacco	46.3
Basic materials	9.0	Basic materials	14.6
Total incl. others	**448.3**	Total incl. others	**591.0**

Main export destinations		Main origins of imports	
	% of total		*% of total*
United States	13.1	Germany	13.3
France	11.8	France	8.4
Germany	11.4	United States	8.0
Ireland	7.2	Netherlands	7.0
Netherlands	6.8	Belgium-Luxembourg	5.7
EU25	62.8	EU25	58.2

Balance of payments, reserves and aid, $bn

Visible exports fob	449.5	Overall balance	-1.3
Visible imports fob	-592.4	Change in reserves	3.4
Trade balance	-142.9	Level of reserves	
Invisibles inflows	675.2	end Dec.	47.0
Invisibles outflows	-587.9	No. months of import cover	0.5
Net transfers	-22.0	Official gold holdings, m oz	10.0
Current account balance	-77.6	Aid given	12.46
– as % of GDP	-3.3	– as % of GDP	0.52
Capital balance	50.3		

Health and education

Health spending, % of GDP	8.2	Education spending, % of GDP	5.6
Doctors per 1,000 pop.	2.1	Enrolment, %: primary	105
Hospital beds per 1,000 pop.	3.9	secondary	98
Improved-water source access,		tertiary	59
% of pop.	100		

Society

No. of households	26.2m	Colour TVs per 100 households	99.5
Av. no. per household	2.3	Telephone lines per 100 pop.	56.2
Marriages per 1,000 pop.	5.2	Mobile telephone subscribers	
Divorces per 1,000 pop.	2.9	per 100 pop.	116.6
Cost of living, Dec. 2007		Computers per 100 pop.	75.8
New York = 100	139	Internet hosts per 1,000 pop.	129.2

UNITED STATES

Area	9,372,610 sq km	Capital	Washington DC
Arable as % of total land	19	Currency	US dollar ($)

People

Population	301.0m	Life expectancy:	men	75.6 yrs
Pop. per sq km	32.1		women	80.8 yrs
Av. ann. growth		Adult literacy		...
in pop. 2010–15	0.89%	Fertility rate (per woman)		2.0
Pop. under 15	20.8%	Urban population		80.8%
Pop. over 60	16.6%			per 1,000 pop.
No. of men per 100 women	97	Crude birth rate		14.0
Human Development Index	95.1	Crude death rate		8.2

The economy

GDP	$13,164	GDP per head	$43,730
Av. ann. growth in real		GDP per head in purchasing	
GDP 1997–2007	3.2%	power parity (USA=100)	100
		Economic freedom index	80.6

Origins of GDP		Components of GDP	
	% of total		% of total
Agriculture	0.9	Private consumption	69.9
Industry, of which:	20.9	Public consumption	19.1
manufacturing	12.6	Non-government investment	16.4
Services[a]	78.2	Exports	11.1
		Imports	-16.9

Structure of employment

	% of total		% of labour force
Agriculture	2	Unemployed 2006	4.6
Industry	20	Av. ann. rate 1995–2006	5.0
Services	78		

Energy

	m TOE		
Total output	1,630.7	Net energy imports as %	
Total consumption	2,340.3	of energy use	30
Consumption per head,			
kg oil equivalent	7,893		

Inflation and finance

Consumer price		av. ann. increase 2001–06	
inflation 2007	2.9%	Narrow money	1.3%
Av. ann. inflation 2002–07	2.9%	Broad money	5.7%
Treasury bill rate, 2007	4.41%	Household saving rate, 2007	0.7%

Exchange rates

	end 2007		December 2007
$ per SDR	1.58	Effective rates	2000 = 100
$ per €	1.47	– nominal	82.40
		– real	85.60

Trade

Principal exports	$bn fob
Capital goods, excl. vehicles	445.9
Industrial supplies	315.5
Consumer goods, excl. vehicles	146.4
Vehicles & products	120.9
Food & beverages	84.2
Total incl. others	**1,163.2**

Principal imports	$bn fob
Industrial supplies	630.7
Consumer goods, excl. vehicles	474.9
Capital goods, excl. vehicles	444.7
Vehicles & products	258.9
Food & beverages	81.7
Total incl. others	**1,953.3**

Main export destinations	% of total
Canada	21.4
Mexico	11.7
China	5.6
Japan	5.4
Germany	4.3
United Kingdom	4.3

Main origins of imports	% of total
China	16.5
Canada	16.0
Mexico	10.8
Japan	7.4
Germany	4.8
United Kingdom	2.9

Balance of payments, reserves and aid, $bn

Visible exports fob	1,026.9	Overall balance	-2.4
Visible imports fob	-1,861.4	Change in reserves	32.8
Trade balance	-834.6	Level of reserves	
Invisibles inflows	1,069.3	end Dec.	221.1
Invisibles outflows	-956.6	No. months of import cover	0.9
Net transfers	-89.6	Official gold holdings, m oz	261.5
Current account balance	-811.5	Aid given	23.53
– as % of GDP	-6.2	– as % of GDP	0.18
Capital balance	826.9		

Health and education

Health spending, % of GDP	15.9	Education spending, % of GDP	5.3
Doctors per 1,000 pop.	3.0	Enrolment, %: primary	98
Hospital beds per 1,000 pop.	3.3	secondary	94
Improved-water source access,		tertiary	82
% of pop.	100		

Society

No. of households	113.9m	Colour TVs per 100 households	98.5
Av. no. per household	2.6	Telephone lines per 100 pop.	57.2
Marriages per 1,000 pop.	7.6	Mobile telephone subscribers	
Divorces per 1,000 pop.	3.6	per 100 pop.	77.4
Cost of living, Dec. 2007		Computers per 100 pop.	76.2
New York = 100	100	Internet hosts per 1,000 pop.[b]	967.5

a Including utilities.
b Includes all hosts ending ".com", ".net" and ".org" which exaggerates the numbers.

VENEZUELA

Area	912,050 sq km	Capital	Caracas
Arable as % of total land	3	Currency	Bolivar (Bs)

People

Population	27.2m	Life expectancy:	men	70.9 yrs
Pop. per sq km	29.8		women	76.8 yrs
Av. ann. growth		Adult literacy		93.0%
in pop. 2010–15	1.49%	Fertility rate (per woman)		2.4
Pop. under 15	31.3%	Urban population		88.1%
Pop. over 60	7.5%			per 1,000 pop.
No. of men per 100 women	101	Crude birth rate		22
Human Development Index	79.2	Crude death rate		5.1

The economy

GDP	Bs390trn	GDP per head	$6,690
GDP	$182bn	GDP per head in purchasing	
Av. ann. growth in real		power parity (USA=100)	25.2
GDP 1996–2006	3.2%	Economic freedom index	45.0

Origins of GDP		Components of GDP	
	% of total		% of total
Agriculture	3.8	Private consumption	47.8
Industry, of which:	38.4	Public consumption	11.9
manufacturing	16.5	Investment	23.7
Services	57.8	Exports	31.0
		Imports	-24.7

Structure of employment

	% of total		% of labour force
Agriculture	11	Unemployed 2003	16.8
Industry	20	Av. ann. rate 1995–2003	13.2
Services	69		

Energy

	m TOE		
Total output	204.7	Net energy imports as %	
Total consumption	60.9	of energy use	-236
Consumption per head,			
kg oil equivalent	2,293		

Inflation and finance

			av. ann. increase 2001–06
Consumer price			
inflation 2007	18.7%	Narrow money	63.8%
Av. ann. inflation 2002–07	20.1%	Broad money	48.5%
Money market rate, 2007	8.72%		

Exchange rates

	end 2007		December 2007
Bs per $	2.15	Effective rates	2000 = 100
Bs per SDR	3.39	– nominal	28.80
Bs per €	3.17	– real	86.70

Trade

Principal exports	$bn fob	Principal imports	$bn fob
Oil	62.6	Intermediate goods	20.5
Non-oil	6.6	Capital goods	14.3
		Consumer goods	11.3
Total incl. others	**69.2**	Total incl. others	**46.1**

Main export destinations	% of total	Main origins of imports	% of total
United States	53.5	United States	29.2
Netherlands Antilles	8.8	Colombia	9.6
China	3.7	Brazil	7.9
Spain	3.0	Mexico	6.1

Balance of payments, reserves and debt, $bn

Visible exports fob	65.2	Change in reserves	6.9
Visible imports fob	-32.5	Level of reserves	
Trade balance	32.7	end Dec.	36.7
Invisibles inflows	9.5	No. months of import cover	9.3
Invisibles outflows	-15.0	Official gold holdings, m oz	11.5
Net transfers	-0.0	Foreign debt	44.6
Current account balance	27.1	– as % of GDP	34
– as % of GDP	14.9	– as % of total exports	83
Capital balance	-19.2	Debt service ratio	13
Overall balance	5.1		

Health and education

Health spending, % of GDP	4.7	Education spending, % of GDP	...
Doctors per 1,000 pop.	1.4	Enrolment, %: primary	104
Hospital beds per 1,000 pop.	0.9	secondary	77
Improved-water source access,		tertiary	52
% of pop.	83		

Society

No. of households	5.9m	Colour TVs per 100 households	86.5
Av. no. per household	4.5	Telephone lines per 100 pop.	15.8
Marriages per 1,000 pop.	2.7	Mobile telephone subscribers	
Divorces per 1,000 pop.	0.8	per 100 pop.	69.0
Cost of living, Dec. 2007		Computers per 100 pop.	9.3
New York = 100	38	Internet hosts per 1,000 pop.	5.3

VIETNAM

Area	331,114 sq km	Capital	Hanoi
Arable as % of total land	21	Currency	Dong (D)

People

Population	85.3m	Life expectancy:	men	72.3 yrs
Pop. per sq km	257.6		women	76.2 yrs
Av. ann. growth		Adult literacy		...
in pop. 2010–15	1.20%	Fertility rate (per woman)		2.0
Pop. under 15	29.6%	Urban population		26.7%
Pop. over 60	7.6%			per 1,000 pop.
No. of men per 100 women	100	Crude birth rate		19
Human Development Index	73.3	Crude death rate		5.1

The economy

GDP	D974trn	GDP per head	$720
GDP	$61.0bn	GDP per head in purchasing	
Av. ann. growth in real		power parity (USA=100)	5.4
GDP 1996–2006	8.0%	Economic freedom index	49.8

Origins of GDP		**Components of GDP**	
	% of total		% of total
Agriculture	20.4	Private consumption	63.5
Industry	41.5	Public consumption	6.2
Services	38.1	Investment	32.9
		Exports	69.4
		Imports	-73.5

Structure of employment

	% of total		% of labour force
Agriculture	58	Unemployed 2004	2.1
Industry	17	Av. ann. rate 2003–2004	2.2
Services	25		

Energy

	m TOE		
Total output	69.5	Net energy imports as %	
Total consumption	51.3	of energy use	-36
Consumption per head,			
kg oil equivalent	617		

Inflation and finance

Consumer price		av. ann. increase 2001–06	
inflation 2007	8.3%	Narrow money (M1)	9.1
Av. ann. inflation 2002–07	7.0%	Broad money	22.8
Treasury bill rate, 2006	4.73%		

Exchange rates

	end 2007		December 2007
D per $	16,016	Effective rates	2000 = 100
D per SDR	25,310	– nominal	...
D per €	23,417	– real	...

Trade

Principal exports		Principal imports	
	$bn fob		*$bn cif*
Crude oil	8.1	Machinery & equipment	6.6
Textiles & garments	5.8	Petroleum products	6.2
Footwear	3.5	Steel	3.0
Fisheries products	3.3	Textiles	2.0
Total incl. others	**39.0**	Total incl. others	**44.0**

Main export destinations		Main origins of imports	
	% of total		*% of total*
United States	21.6	China	17.4
Japan	13.5	Singapore	12.7
Australia	9.6	Japan	9.5
China	5.8	South Korea	9.0
Singapore	5.2	Thailand	7.2
Germany	4.9	Malaysia	4.1
United Kingdom	3.8	Hong Kong	3.5

Balance of payments, reserves and debt, $bn

Visible exports fob	39.8	Change in reserves	4.4
Visible imports fob	-42.6	Level of reserves	
Trade balance	-2.8	end Dec.	13.6
Invisibles inflows	5.8	No. months of import cover	3.3
Invisibles outflows	-7.2	Official gold holdings, m oz	0.0
Net transfers	4.0	Foreign debt	20.2
Current account balance	-0.2	– as % of GDP	34
– as % of GDP	-0.3	– as % of total exports	83
Capital balance	4.5	Debt service ratio	...
Overall balance	4.3		

Health and education

Health spending, % of GDP	6.0	Education spending, % of GDP	...
Doctors per 1,000 pop.	0.7	Enrolment, %: primary	...
Hospital beds per 1,000 pop.	1.4	secondary	...
Improved-water source access,		tertiary	...
% of pop.	85		

Society

No. of households	25.6m	Colour TVs per 100 households	70.8
Av. no. per household	3.3	Telephone lines per 100 pop.	32.2
Marriages per 1,000 pop.	12.1	Mobile telephone subscribers	
Divorces per 1,000 pop.	0.5	per 100 pop.	18.2
Cost of living, Dec. 2007		Computers per 100 pop.	1.4
New York = 100	61	Internet hosts per 1,000 pop.	1.4

ZIMBABWE

Area	390,759 sq km	Capital	Harare
Arable as % of total land	8	Currency	Zimbabwe dollar (Z$)

People

Population	13.1m	Life expectancy: men		44.1 yrs
Pop. per sq km	33.5		women	42.7 yrs
Av. ann. growth		Adult literacy		89.4%
in pop. 2010–15	1.06%	Fertility rate (per woman)		2.9
Pop. under 15	45.7%	Urban population		35.9%
Pop. over 60	5.2%			per 1,000 pop.
No. of men per 100 women	99	Crude birth rate		31
Human Development Index	51.3	Crude death rate		17.9

The economy

GDP[a]	Z$76,441bn	GDP per head	$260
GDP[a]	$3.4bn	GDP per head in purchasing	
Av. ann. growth in real		power parity (USA=100)	4.5
GDP 1995–2005	-2.6%	Economic freedom index	29.8

Origins of GDP[a]

	% of total
Agriculture	18
Industry, of which:	23
manufacturing	13
Services	59

Components of GDP[a]

	% of total
Private consumption	69.6
Public consumption	26.6
Investment	17.3
Exports	42.8
Imports	-52.9

Structure of employment

	% of total		% of labour force
Agriculture	...	Unemployed 2002	8.2
Industry	...	Av. ann. rate 1997–2002	7.1
Services	...		

Energy

	m TOE		
Total output	8.9	Net energy imports as %	
Total consumption	9.7	of energy use	9
Consumption per head,			
kg oil equivalent	714		

Inflation and finance

Consumer price			av. ann. increase 2001–06
inflation 2007	24,411%	Narrow money (M1)	445%
Av. ann. inflation 2002–07	1,091.2%	Broad money	450%
Deposit rate, 2006	203.4%		

Exchange rates

	end 2007		December 2007
Z$ per $	...	Effective rates	2000 = 100
Z$ per SDR	...	– nominal	...
Z$ per €	...	– real	...

Trade

Principal exports[bc]		Principal imports[bc]	
	$m fob		*$m cif*
Gold	276	Fuels	413
Tobacco	235	Chemicals	401
Ferro-alloys	185	Machinery & transport equip.	271
Platinum	78	Manufactured products	269
Total incl. others	**1,680**	Total incl. others	**2,009**

Main export destinations		Main origins of imports	
	% of total		*% of total*
South Africa	43.5	South Africa	46.1
China	8.5	China	5.9
Zambia	8.4	Botswana	4.8
Japan	8.0	Zambia	4.1
United States	6.6	Germany	2.9

Balance of payments[b], reserves and debt, $bn

Visible exports fob	1.7	Change in reserves[e]	0.0
Visible imports fob	-2.1	Level of reserves[e]	
Trade balance	-0.5	end Dec.	0.1
Invisibles, net	-0.7	No. months of import cover[e]	0.6
Net transfers	0.3	Official gold holdings, m oz[e]	0.1
Current account balance	-0.5	Foreign debt	4.7
– as % of GDP	-14.5	– as % of GDP	110
Capital balance[d]	-0.4	– as % of total exports	248
Overall balance[d]	-0.4	Debt service ratio[a]	13

Health and education

Health spending, % of GDP	8.1	Education spending, % of GDP	...
Doctors per 1,000 pop.	0.2	Enrolment, %: primary	101
Hospital beds per 1,000 pop.	...	secondary	40
Improved-water source access,		tertiary	4
% of pop.	81		

Society

No. of households	3.3m	Colour TVs per 100 households	...
Av. no. per household	3.9	Telephone lines per 100 pop.	2.6
Marriages per 1,000 pop.	...	Mobile telephone subscribers	
Divorces per 1,000 pop.	...	per 100 pop.	6.5
Cost of living, Dec. 2007		Computers per 100 pop.	6.5
New York = 100	...	Internet hosts per 1,000 pop.	1.4

a 2005
b Estimates.
c 2004
d 2001 estimates.
e 2002

EURO AREA[a]

Area	2,497,000 sq km	Capital	–
Arable as % of total land	25	Currency	Euro (€)

People

Population	311.5m	Life expectancy:	men	76.5 yrs
Pop. per sq km	124.8		women	82.5 yrs
Av. ann. growth		Adult literacy		...
in pop. 2010–15	0.13%	Fertility rate (per woman)		1.5
Pop. under 15	16.0%	Urban population		76.2%
Pop. over 60	22.0%			per 1,000 pop.
No. of men per 100 women	96	Crude birth rate		9.3
Human Development Index	94.2	Crude death rate[b]		9.7

The economy

GDP	€8,429bn	GDP per head	$34,150
GDP	$10,636bn	GDP per head in purchasing	
Av. ann. growth in real		power parity (USA=100)	70.9
GDP 1997–2007	2.1%	Economic freedom index	68.1

Origins of GDP		Components of GDP	
	% of total		% of total
Agriculture	2	Private consumption	58
Industry, of which:	26	Public consumption	21
manufacturing	19	Investment	20
Services	72	Exports	37
		Imports	-36

Structure of employment

	% of total		% of labour force
Agriculture	4.3	Unemployed 2006	8.2
Industry	27.8	Av. ann. rate 1995–2006	9.0
Services	69.5		

Energy

	m TOE		
Total output	450.5	Net energy imports as %	
Total consumption	1,244.4	of energy use	64
Consumption per head,			
kg oil equivalent	3,961		

Inflation and finance

Consumer price		av. ann. increase 2001–06	
inflation 2007	2.1%	Narrow money (M1)	10.5%
Av. ann. inflation 2002–07	2.1%	Broad money	7.4%
Interbank rate, 2007	3.86%	Household saving rate, 2007	9.9%

Exchange rates

	end 2007		December 2007
€ per $	0.68	Effective rates	2000 = 100
€ per SDR	1.07	– nominal	134.81
		– real	128.42

Trade[b]

Principal exports

	$bn fob
Machinery & transport equip.	652
Manufactures	385
Chemicals	237
Energy and raw materials	108
Food, drink & tobacco	75
Total incl. others	**1,754**

Principal imports

	$bn cif
Machinery & transport equip.	513
Energy & raw materials	501
Manufactures	446
Chemicals	138
Food, drink & tobacco	86
Total incl. others	**1,722**

Main export destinations

	% of total
United States	23.2
Switzerland	7.5
Russia	6.3
China	5.5
Japan	4.3

Main origins of imports

	% of total
China	14.4
United States	13.2
Russia	10.4
Norway	5.9
Japan	5.7

Balance of payments, reserves and aid, $bn

Visible exports fob	1,753.9	Overall balance	2.6
Visible imports fob	1,721.8	Change in reserves	51.6
Trade balance	32.1	Level of reserves	
Invisibles inflows	1,119.4	end Dec.	429.2
Invisibles outflows	1,067.3	No. months of import cover	1.8
Net transfers	-96.7	Official gold holdings, m oz	375.9
Current account balance	-12.5	Aid given	40.39
– as % of GDP	-0.1	– as % of GDP	0.38
Capital balance	155.5		

Health and education

Health spending, % of GDP	9.6	Education spending, % of GDP	5.3
Doctors per 1,000 pop.	4.0	Enrolment, %: primary	...
Hospital beds per 1,000 pop.	6.6	secondary	...
Improved-water source access,		tertiary	...
% of pop.	100		

Society

No. of households	130.5	Colour TVs per 100 households	97.0
Av. no. per household	2.39	Telephone lines per 100 pop.	53.2
Marriages per 1,000 pop.	4.6	Mobile telephone subscribers	
Divorces per 1,000 pop.	1.9	per 100 pop.	106.7
Cost of living, Dec. 2007		Computers per 100 pop.	47.6
New York = 100	...	Internet hosts per 1,000 pop.	255.8

---

a Data refer to the 12 EU members that had adopted the euro before December 31 2006: Austria, Belgium, France, Finland, Germany, Greece, Ireland, Italy, Luxembourg, Netherlands, Portugal and Spain.

b EU25 data, excluding intra-trade.

WORLD

| Area | 148,698,382 sq km | Capital | ... |
| Arable as % of total land | 11 | Currency | ... |

People

Population	6,540.3m	Life expectancy: men	65.0 yrs
Pop. per sq km	43.9	women	69.5 yrs
Av. ann. growth		Adult literacy	82.4%
in pop. 2010–15	1.10%	Fertility rate (per woman)	2.5
Pop. under 15	28.3%	Urban population	48.7%
Pop. over 60	10.3%		per 1,000 pop.
No. of men per 100 women	102	Crude birth rate	21
Human Development Index	74.3	Crude death rate	8.6

The economy

GDP	$48.5trn	GDP per head	$7,410
Av. ann. growth in real		GDP per head in purchasing	
GDP 1996–2006	3.4%	power parity (USA=100)	21.0
		Economic freedom index	57.4

Origins of GDP		**Components of GDP**	
	% of total		% of total
Agriculture	3	Private consumption	61
Industry, of which:	28	Public consumption	17
manufacturing	18	Investment	22
Services	69	Exports	27
		Imports	-27

Structure of employment[a]

	% of total		% of labour force
Agriculture	...	Unemployed 2006	6.0
Industry	...	Av. ann. rate 1995–2006	6.6
Services	...		

Energy

	m TOE		
Total output	11,441.1	Net energy imports as %	
Total consumption	11,209.7	of energy use	-2
Consumption per head,			
kg oil equivalent	1,796		

Inflation and finance

Consumer price		av. ann. increase 2001–06	
inflation 2007	3.8%	Narrow money (M1)[a]	8.4%
Av. ann. inflation 2002–07	3.6%	Broad money[a]	6.6%
LIBOR $ rate, 3-month, 2007	5.30%	Household saving rate, 2007[a]	4.3%

Trade

World exports

	$bn fob		$bn fob
Manufactures	7,825	Ores & metals	321
Fuels	1,043	Agricultural raw materials	214
Food	730		
		Total incl. others	**10,685**

Main export destinations

Main origins of imports

	% of total		% of total
United States	15.1	China	9.8
Germany	7.4	Germany	8.8
China	5.9	United States	8.5
France	4.6	Japan	5.6
United Kingdom	4.5	France	4.0
Japan	4.4	United Kingdom	3.2

Balance of payments, reserves and aid, $bn

Visible exports fob	11,978	Overall balance	0
Visible imports fob	-11,845	Change in reserves	947
Trade balance	133	Level of reserves	
Invisibles inflows	5,766	end Dec.	5,643
Invisibles outflows	-5,736	No. months of import cover	4
Net transfers	-3	Official gold holdings, m oz	879.0
Current account balance	160	Aid given[b]	108.4
– as % of GDP	0.3	– as % of GDP[b]	0.30
Capital balance	-80		

Health and education

Health spending, % of GDP	10.1	Education spending, % of GDP	4.6
Doctors per 1,000 pop.	1.5	Enrolment, %: primary	106
Hospital beds per 1,000 pop.	...	secondary	65
Improved-water source access,		tertiary	24
% of pop.	83		

Society

No. of households	...	TVs per 100 households	...
Av. no. per household	...	Telephone lines per 100 pop.	19.6
Marriages per 1,000 pop.	...	Mobile telephone subscribers	
Divorces per 1,000 pop.	...	per 100 pop.	41.6
Cost of living, Dec. 2007		Computers per 100 pop.	10.6
New York = 100	...	Internet hosts per 1,000 pop.	82.8

a OECD countries.
b OECD, non-OECD Europe and Middle East countries.

Glossary

Balance of payments The record of a country's transactions with the rest of the world. The **current account** of the balance of payments consists of: visible trade (goods); "invisible" trade (services and income); private transfer payments (eg, remittances from those working abroad); official transfers (eg, payments to international organisations, famine relief). Visible imports and exports are normally compiled on rather different definitions to those used in the trade statistics (shown in principal imports and exports) and therefore the statistics do not match. The **capital account** consists of long- and short-term transactions relating to a country's assets and liabilities (eg, loans and borrowings). The current account and the capital account, plus an errors and omissions item, make up the **overall balance**. In the country pages of this book this item is included in the overall balance. **Changes in reserves** include gold at market prices and are shown without the practice often followed in balance of payments presentations of reversing the sign.

Big Mac index A light-hearted way of looking at exchange rates. If the dollar price of a burger at McDonald's in any country is higher than the price in the United States, converting at market exchange rates, then that country's currency could be thought to be over-valued against the dollar and vice versa.

Body-mass index A measure for assessing obesity – weight in kilograms divided by height in metres squared. An index of 30 or more is regarded as an indicator of obesity; 25 to 29.9 as over-weight. Guidelines vary for men and for women and may be adjusted for age.

CFA Communauté Financière Africaine. Its members, most of the francophone African nations, share a common currency, the CFA franc, which used to be pegged to the French franc but is now pegged to the euro.

Cif/fob Measures of the value of merchandise trade. Imports include the cost of "carriage, insurance and freight" (cif) from the exporting country to the importing. The value of exports does not include these elements and is recorded "free on board" (fob). Balance of payments statistics are generally adjusted so that both exports and imports are shown fob; the cif elements are included in invisibles.

Crude birth rate The number of live births in a year per 1,000 population. The crude rate will automatically be relatively high if a large proportion of the population is of childbearing age.

Crude death rate The number of deaths in a year per 1,000 population. Also affected by the population's age structure.

Debt, foreign Financial obligations owed by a country to the rest of the world and repayable in foreign currency. **The debt service ratio** is debt service (principal repayments plus interest payments) expressed as a percentage of the country's earnings from exports of goods and services.

EU European Union. Members are: Austria, Belgium, Denmark, Finland, France, Germany, Greece, Ireland, Italy, Luxembourg, Netherlands, Portugal, Spain, Sweden and the United Kingdom and, as of May 1 2004, Cyprus, Czech Republic, Estonia, Hungary, Latvia, Lithuania, Malta, Poland, Slovakia and Slovenia and, as of January 1 2007, Bulgaria and Romania.

Effective exchange rate The nominal index measures a currency's depreciation (figures below 100) or appreciation (figures over 100) from a base date against a trade-weighted basket of the currencies of the country's main trading partners. The real effective exchange rate reflects adjustments for relative movements in prices or costs.

Euro area The 15 euro area members of the EU are Austria, Belgium, Finland, France, Germany, Greece, Ireland, Italy, Luxembourg, Netherlands, Portugal and Spain and, from January 1 2007,

Slovenia. Cyprus and Malta joined on January 1 2008. Their common currency is the euro, which came into circulation on January 1 2002.

Fertility rate The average number of children born to a woman who completes her childbearing years.

G7 Group of seven countries: United States, Japan, Germany, United Kingdom, France, Italy and Canada.

GDP Gross domestic product. The sum of all output produced by economic activity within a country. GNP (gross national product) and GNI (gross national income) include net income from abroad eg, rent, profits.

Household saving rate Household savings as % of disposable household income.

Import cover The number of months of imports covered by reserves ie, reserves ÷ $\frac{1}{12}$ annual imports (visibles and invisibles).

Inflation The annual rate at which prices are increasing. The most common measure and the one shown here is the increase in the consumer price index.

Internet hosts Websites and other computers that sit permanently on the internet.

Life expectancy The average length of time a baby born today can expect to live.

Literacy is defined by UNESCO as the ability to read and write a simple sentence, but definitions can vary from country to country.

Median age Divides the age distribution into two halves. Half of the population is above and half below the median age.

Money supply A measure of the "money" available to buy goods and services. Various definitions exist. The measures shown here are based on definitions used by the IMF and may differ from measures used nationally. Narrow money (M1)

consists of cash in circulation and demand deposits (bank deposits that can be withdrawn on demand). "Quasi-money" (time, savings and foreign currency deposits) is added to this to create broad money.

OECD Organisation for Economic Co-operation and Development. The "rich countries" club was established in 1961 to promote economic growth and the expansion of world trade. It is based in Paris and now has 30 members.

Opec Organisation of Petroleum Exporting Countries. Set up in 1960 and based in Vienna, Opec is mainly concerned with oil pricing and production issues. Members are; Algeria, Indonesia, Iran, Iraq, Kuwait, Libya, Nigeria, Qatar, Saudi Arabia, United Arab Emirates and Venezuela.

PPP Purchasing power parity. PPP statistics adjust for cost of living differences by replacing normal exchange rates with rates designed to equalise the prices of a standard "basket"of goods and services. These are used to obtain PPP estimates of GDP per head. PPP estimates are shown on an index, taking the United States as 100.

Real terms Figures adjusted to exclude the effect of inflation.

Reserves The stock of gold and foreign currency held by a country to finance any calls that may be made for the settlement of foreign debt.

SDR Special drawing right. The reserve currency, introduced by the IMF in 1970, was intended to replace gold and national currencies in settling international transactions. The IMF uses SDRs for book-keeping purposes and issues them to member countries. Their value is based on a basket of the US dollar (with a weight of 44%), the euro (34%), the Japanese yen (11%) and the pound sterling (11%).

List of countries

Whenever data is available, the world rankings consider 186 countries: all those which had (in 2006) or have recently had a population of at least 1m or a GDP of at least $1bn. Here is a list of them.

	Population	GDP	GDP per head	Area '000 sq	Median age
	m, 2006	$bn, 2006	$PPP, 2006	km	yrs, 2007
Afghanistan	31.1	8.4	1,000	652	16.7
Albania	3.1	9.1	5,890	29	28.3
Algeria	33.4	114.7	6,350	2,382	24.0
Andorra	0.07	2.8ab	38,800ab	0.4	42.0
Angola	16.4	45.2	4,430	1,247	16.7
Argentina	39.1	214.2	11,990	2,767	28.9
Armenia	3.0	6.4	4,880	30	31.7
Aruba	0.10	2.3ab	21,800ab	0.2	37.0
Australia	20.4	780.5	35,550	7,682	36.6
Austria	8.2	322.0	36,050	84	40.6
Azerbaijan	8.5	19.9	6,280	87	27.5
Bahamas	0.31	6.2a	23,070ab	14	27.6
Bahrain	0.71	16.0b	33,450b	1	29.8
Bangladesh	144.4	61.9	1,160	144	22.1
Barbados	0.28	3.4	15,920b	0.4	34.7
Belarus	9.7	36.9	9,730	208	37.8
Belgium	10.4	394.0	33,540	31	40.6
Belize	0.29	1.2	7,850	23	21.2
Benin	8.7	4.8	1,260	113	17.6
Bermuda	0.07	4.5ab	69,900ab	1	39.0
Bhutan	2.2	0.9	4,010	47	20.1
Bolivia	9.4	11.2	3,940	1,099	20.8
Bosnia	3.9	12.3	6,490	51	38.0
Botswana	1.8	10.6	12,510	581	19.9
Brazil	188.9	1,067.5	8,950	8,512	26.8
Brunei	0.37	11.6	49,900	6	26.2
Bulgaria	7.7	31.5	10,270	111	40.6
Burkina Faso	13.6	6.2	1,130	274	16.2
Burundi	7.8	0.9	330	28	17.0
Cambodia	14.4	7.3	1,620	181	20.3
Cameroon	16.6	18.3	2,090	475	18.8
Canada	32.6	1,271.6	36,710	9,971	38.6
Cape Verde	0.42	1.1	2,700	4	19.3
Cayman Islands	0.05	1.9ab	43,800ab	0.3	35.0
Central African Rep	4.1	1.5	690	622	18.1
Chad	10.0	6.5	1,480	1,284	16.3
Channel Islands	0.10	7.8ab	51,820ab	0.2	39.7
Chile	16.5	145.8	13,030	757	30.6
China	1,323.6	2,644.7	4,640	9,561	32.6
Colombia	46.3	153.4	6,380	1,142	25.4

	Population m, 2006	GDP $bn, 2006	GDP per head $PPP, 2006	Area '000 sq km	Median age yrs, 2007
Congo-Brazzaville	4.1	7.4	3,490	342	16.3
Congo-Kinshasa	59.3	8.5	280	2,345	16.3
Costa Rica	4.4	22.2	9,560	51	26.1
Côte d'Ivoire	18.5	17.6	1,650	322	18.5
Croatia	4.6	42.9	14,310	57	40.6
Cuba	11.3	40.0[a]	3,900[a]	111	35.6
Cyprus	0.78	18.4	25,880	9	35.3
Czech Republic	10.2	143.0	22,120	79	39.0
Denmark	5.4	275.4	35,690	43	39.5
Dominican Republic	9.0	31.8	5,870	48	23.3
Ecuador	13.4	41.4	7,150	272	24.0
Egypt	75.4	107.5	4,950	1,000	22.8
El Salvador	7.0	18.7	5,770	21	23.3
Equatorial Guinea	0.54	8.6	27,160	28	17.6
Eritrea	4.6	1.1	680	117	17.4
Estonia	1.3	16.4	18,970	45	38.9
Ethiopia	79.3	13.3	640	1,134	17.5
Faroe Islands	0.05	1.7[ab]	31,000[ab]	1	34.0
Fiji	0.91	3.1	4,550	18	24.5
Finland	5.3	210.7	33,020	338	40.9
France	60.7	2,248.1[c]	31,990	544	39.3
French Polynesia	0.28	3.8[ab]	17,500[ab]	3	26.9
Gabon	1.4	9.5	14,210	268	19.4
Gambia, The	1.6	0.5	1,130	11	19.8
Georgia	4.4	7.7	4,010	70	35.5
Germany	82.7	2,896.9	32,320	358	42.1
Ghana	22.6	12.9	1,250	239	19.8
Greece	11.1	308.4	31,380	132	39.7
Greenland	0.06	1.7[ab]	20,000[ab]	2,176	34.0
Guadeloupe	0.45	9.6[b]	19,660[b]	2	34.1
Guam	0.17	2.5[ab]	15,000[ab]	1	28.1
Guatemala	12.9	35.3	5,180	109	18.1
Guinea	9.6	3.3	1,150	246	18.0
Guinea-Bissau	1.6	0.3	480	36	16.2
Haiti	8.6	5.0	1,220	28	20.0
Honduras	7.4	9.2	3,540	112	19.8
Hong Kong	7.1	189.8	39,060	1	38.9
Hungary	10.1	112.9	18,280	93	38.8
Iceland	0.30	16.3	36,920	103	34.1
India	1,119.5	911.8	2,470	3,287	24.3
Indonesia	225.5	364.8	3,450	1,904	26.5

	Population	GDP	GDP per head	Area '000 sq	Median age
	m, 2006	$bn, 2006	$PPP, 2006	km	yrs, 2007
Iran	70.3	217.9	9,910	1,648	23.4
Iraq	29.6	42.1[a]	1,900[a]	438	19.1
Ireland	4.2	220.1	40,270	70	34.2
Israel	6.8	140.5	24,100	21	28.9
Italy	58.1	1,851.0	29,050	301	42.3
Jamaica	2.7	10.0	7,570	11	24.9
Japan	128.2	4,368.4	31,950	378	42.9
Jordan	5.8	14.1	4,630	89	21.3
Kazakhstan	14.8	81.0	9,830	2,717	29.4
Kenya	35.1	22.8	1,470	583	17.9
Kuwait	2.8	102.1	43,550[b]	18	29.5
Kyrgyzstan	5.3	2.8	1,810	199	23.8
Laos	6.1	3.4	1,980	237	19.1
Latvia	2.3	20.1	15,350	64	39.5
Lebanon	3.6	22.7	9,740	10	26.8
Lesotho	1.8	1.5	1,440	30	19.2
Liberia	3.4	0.6	330	111	16.3
Libya	6.0	50.3	11,620	1,760	23.9
Lithuania	3.4	29.8	15,740	65	37.8
Luxembourg	0.47	41.5	75,610	3	38.1
Macau	0.45	14.2	43,950	0.02	36.6
Macedonia	2.0	6.2	7,850	26	34.2
Madagascar	19.1	5.5	880	587	17.8
Malawi	13.2	3.2	700	118	16.3
Malaysia	25.8	150.7	12,540	333	24.7
Mali	13.9	5.9	1,060	1,240	15.8
Malta	0.4	6.4	21,720	0.3	38.1
Martinique	0.40	9.1[b]	21,050[b]	1	36.4
Mauritania	3.2	2.7	1,890	1,031	18.4
Mauritius	1.3	6.3	10,570	2	30.4
Mexico	108.3	839.2	12,180	1,973	25.0
Moldova	4.2	3.4	2,380	34	33.0
Mongolia	2.7	3.1	2,890	1,565	23.7
Montenegro	0.69	2.5	9,030	14	35.0
Morocco	31.9	65.4	3,920	447	24.2
Mozambique	20.2	6.8	740	799	17.7
Myanmar	51.0	12.0[b]	840[b]	677	25.5
Namibia	2.1	6.6	4,820	824	18.6
Nepal	27.7	8.9	1,000	147	20.1
Netherlands	16.4	662.3	36,560	42	39.3
Netherlands Antilles	0.22	2.8[ab]	16,000[ab]	1	36.2
New Caledonia	0.22	3.3[ab]	15,000[ab]	19	28.4

	Population	GDP	GDP per head	Area '000 sq	Median age
	m, 2006	$bn, 2006	$PPP, 2006	km	yrs, 2007
New Zealand	4.1	104.5	25,520	271	35.8
Nicaragua	5.6	5.3	2,790	130	19.7
Niger	14.4	3.7	630	1,267	15.5
Nigeria	134.4	115.3	1,610	924	17.5
North Korea	22.6	40.0[ab]	1,800[a]	121	31.1
Norway	4.6	334.9	50,080	324	38.2
Oman	2.6	30.8[b]	20,350[b]	310	22.3
Pakistan	161.2	126.8	2,360	804	20.0
Panama	3.3	17.1	9,250	77	26.1
Papua New Guinea	6.0	5.7	1,820	463	19.7
Paraguay	6.3	9.3	4,030	407	20.8
Peru	28.4	92.4	7,090	1,285	24.2
Philippines	84.5	117.6	3,150	300	22.2
Poland	38.5	338.7	14,840	313	36.5
Portugal	10.5	194.7	20,780	89	39.5
Puerto Rico	4.0	74.9[a]	19,100[a]	9	33.3
Qatar	0.89	52.7	70,720[b]	11	30.9
Réunion	0.80	14.6[b]	17,160[b]	3	29.3
Romania	21.6	121.6	10,430	238	36.7
Russia	142.5	986.9	13,120	17,075	37.3
Rwanda	9.2	2.5	740	26	17.5
Saudi Arabia	25.2	349.1	22,300	2,200	21.6
Senegal	11.9	9.2	1,590	197	18.2
Serbia	10.5	32.0	9,430	88	36.5
Sierra Leone	5.7	1.5	630	72	18.4
Singapore	4.4	132.2	44,710	1	37.5
Slovakia	5.4	55.0	17,730	49	35.6
Slovenia	2.0	37.3	24,360	20	40.2
Somalia	8.5	2.5[a]	600[a]	638	17.9
South Africa	47.6	255.2	9,090	1,226	23.5
South Korea	48.0	888.0	22,990	99	35.1
Spain	43.4	1,224.7	28,650	505	38.6
Sri Lanka	20.9	27.0	3,750	66	29.6
Sudan	37.0	37.4	1,930	2,506	20.1
Suriname	0.47	2.1	7,890	164	25.1
Swaziland	1.0	2.6	4,670	17	18.1
Sweden	9.1	383.8	34,190	450	40.1
Switzerland	7.3	380.4	37,190	41	40.8
Syria	19.5	33.4	4,230	185	20.6
Taiwan	22.8	364.6	31,790[a]	36	35.0
Tajikistan	6.6	2.8	1,610	143	19.3

	Population	GDP	GDP per head	Area '000 sq	Median age
	m, 2006	$bn, 2006	$PPP, 2006	km	yrs, 2007
Tanzania	39.0	12.8	1,000	945	18.2
Thailand	64.8	206.3	7,600	513	30.5
Timor-Leste	1.0	0.4	2,140	15	18.4
Togo	6.3	2.2	780	57	17.9
Trinidad & Tobago	1.3	18.1	17,720	5	29.4
Tunisia	10.2	30.3	6,860	164	26.8
Turkey	74.2	402.7	8,420	779	26.3
Turkmenistan	4.9	10.5	4,290[b]	488	23.3
Uganda	29.9	9.4	890	241	14.8
Ukraine	46.0	106.5	6,210	604	39.0
United Arab Emirates	4.7	129.7[b]	33,480[b]	84	29.0
United Kingdom	59.8	2,377.0	33,090	243	39.0
United States	301	13,163.9	43,970	9,373	36.1
Uruguay	3.5	19.3	10,200	176	32.1
Uzbekistan	27.0	17.2	2,190	447	22.6
Venezuela	27.2	181.9	11,060	912	24.7
Vietnam	85.3	61.0	2,360	331	24.9
Virgin Islands (US)	0.11	1.6[ab]	14,500[ab]	0.4	35.0
West Bank and Gaza	3.8	4.1	3,600	6	17.1
Yemen	21.6	19.1	2,260	528	16.5
Zambia	11.9	10.7	1,260	753	16.7
Zimbabwe	13.1	3.4[b]	2,000[a]	391	18.7
Euro area (12)	311.5	10,636.4	31,180	2,497	40.1
World	6,540.3	48,461.9	9,250	148,698	28.1

a Estimate.
b Latest available year.
c Including French Guiana, Guadeloupe, Martinique and Réunion.

Sources

Academy of Motion Picture Arts and Sciences
Airports Council International, *Worldwide Airport Traffic Report*
Amnesty International
ASEAN, Asian Development Report

BP, *Statistical Review of World Energy*
British Mountaineering Council
Business Software Alliance

CB Richard Ellis, *Global Market Rents*
Central banks
Central Intelligence Agency, *The World Factbook*
Confederation of Swedish Enterprise
Corporate Resources Group, *Quality of Living Report*
Council of Europe

The Economist
www.economist.com
Economist Intelligence Unit, *Cost of Living Survey*; *Country Forecasts*; *Country Reports*; *E-readiness rankings*; *Global Outlook – Business Environment Rankings*
ERC Statistics International, *World Cigarette Report*
Euromonitor, *International Marketing Data and Statistics*; *European Marketing Data and Statistics*
Europa Publications, *The Europa World Yearbook*
Eurostat, *Statistics in Focus*

Financial Times Business Information, *The Banker*
Food and Agriculture Organisation

The Heritage Foundation, *Index of Economic Freedom*
Human Rights Research

IFPI
IMD, *World Competitiveness Yearbook*
IMF, *Direction of Trade*; *International Financial Statistics*; *World Economic Outlook*
International Centre for Prison Studies, *World prison Brief*
International Cocoa Organisation, *Quarterly Bulletin of Cocoa Statistics*
International Coffee Organisation
International Cotton Advisory Committee, *Bulletin*
International Grains Council, *The Grain Market Report*
International Institute for Strategic Studies, *Military Balance*
International Labour Organisation
International Obesity Task Force
International Road Federation, *World Road Statistics*
International Rubber Study Group, *Rubber Statistical Bulletin*
International Sugar Organisation, *Statistical Bulletin*
International Tea Committee, *Annual Bulletin of Statistics*
International Telecommunication Union, *ITU Indicators*
International Union of Railways
ISTA Mielke, *Oil World*

Johnson Matthey

Morgan Stanley Research

National statistics offices
Network Wizards
Nobel Foundation

OECD, *Development Assistance Committee Report*; *Economic Outlook*; *Environmental Data*; *Programme for International Student Assessment*

Population Reference Bureau
Privacy International

Reporters Without Borders, *Press Freedom Index*

Space.com
Standard & Poor's *Emerging Stock Markets Factbook*

Taiwan Statistical Data Book
The Times, *Atlas of the World*
Time Inc Magazines, *Fortune International*
Transparency International

UN, *Demographic Yearbook*; *Global Refugee Trends*; *Review of Maritime Transport*; *State of World Population Report*; *Statistical Chart on World Families*; *Survey on Crime Trends*; *Trends in Total Migrant Stock*; *World Contraceptive Use*; *World Population Prospects*; *World Urbanisation Prospects*
UNAIDS, *Report on the Global AIDS Epidemic*
UNCTAD, *Review of Maritime Transport*; *World Investment Report*
UNCTAD/WTO International Trade Centre
UN Development Programme, *Human Development Report*

UNESCO, website: unescostat. unesco.org
Unicef, *Child Poverty in Perspective*
Union Internationale des Chemins de Fer, *Statistiques Internationales des Chemins de Fer*
US Census Bureau
US Department of Agriculture
University of Michigan, Windows to the Universe website

WHO, *World Health Statistics Annual*; *World Report on Violence and Health*; *Health Behaviour in School-aged Children*
The Woolmark Company
World Bank, *Doing Business*; *Global Development Finance*; *World Development Indicators*; *World Development Report*
World Bureau of Metal Statistics, *World Metal Statistics*
World Economic Forum/Harvard University, *Global Competitiveness Report*
World Resources Institute, *World Resources*
World Tourism Organisation, *Yearbook of Tourism Statistics*
World Trade Organisation, *Annual Report*

$$
\begin{array}{r}
3.2 \\
3.4\,\overline{)110.} \\
102 \\
\hline
80 \\
\end{array}
$$

3.

$$
7\,6 \div \begin{array}{c}20 \end{array}
$$

$$
\begin{array}{r}
4.3\,\overline{)3\,8.0} \\
301 \\
\hline
800 \\
\end{array}
$$

$$
\begin{array}{r}
2.64 \\
34\,\overline{)90.0} \\
68 \\
\hline
220 \\
204 \\
\hline
160 \\
136 \\
\end{array}
$$

$$
\begin{array}{r}
7.6 \\
3.2 \\
\hline
4.4 \\
\end{array}
$$

$$
4.4\,\overline{)17.6}
$$

$$
\begin{array}{r}
58 \\
7.6\,\overline{)4.4^{0}} \\
380 \\
\hline
600 \\
\end{array}
$$

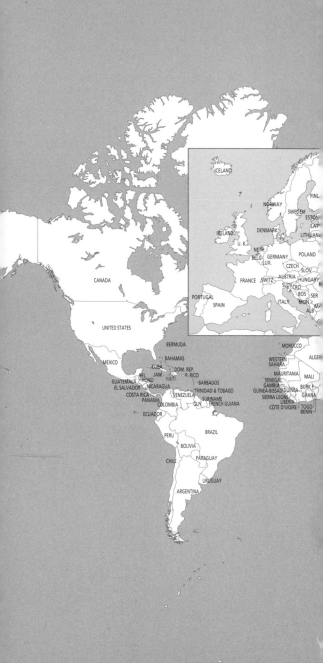